Talking Chicken

Talking Chicken

Practical Advice On Heirloom Chickens & Eggs

✦ Selection ✦ Breeding
✦ Raising ✦ Marketing

• KELLY KLOBER •

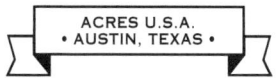

ACRES U.S.A.
• AUSTIN, TEXAS •

TALKING CHICKEN

Copyright © 2011 Kelly Klober

Acres U.S.A.
P.O. Box 91299
Austin, Texas 78709 U.S.A.
(512) 892-4400 • fax (512) 892-4448
info@acresusa.com • www.acresusa.com

Printed in the United States of America

Publisher's Cataloging-in-Publication

Cover image of Dominique cockerel and pullet courtesy of Bryan K. Oliver.

Klober, Kelly, 1949-
Talking chicken / Kelly Klober. Austin, TX, ACRES U.S.A., 2011
 x, 398 pp., 23 cm.
 Includes Index
 Includes Bibliography
 Includes Illustrations
 ISBN 978-1-60173-021-3 (trade)

 1. Animal husbandry — Poultry. 2. Poultry — feeding and feeds.
 3. Poultry breeds. 4. Poultry hatching. 5. Organic farming.
 6. Poultry folklore. I. Klober, Kelly, 1949- II. Title.

 SF487.K56 2011   636.5

To my wife, Phyllis, you have always been
my helpmate and inspiration.

• CONTENTS •

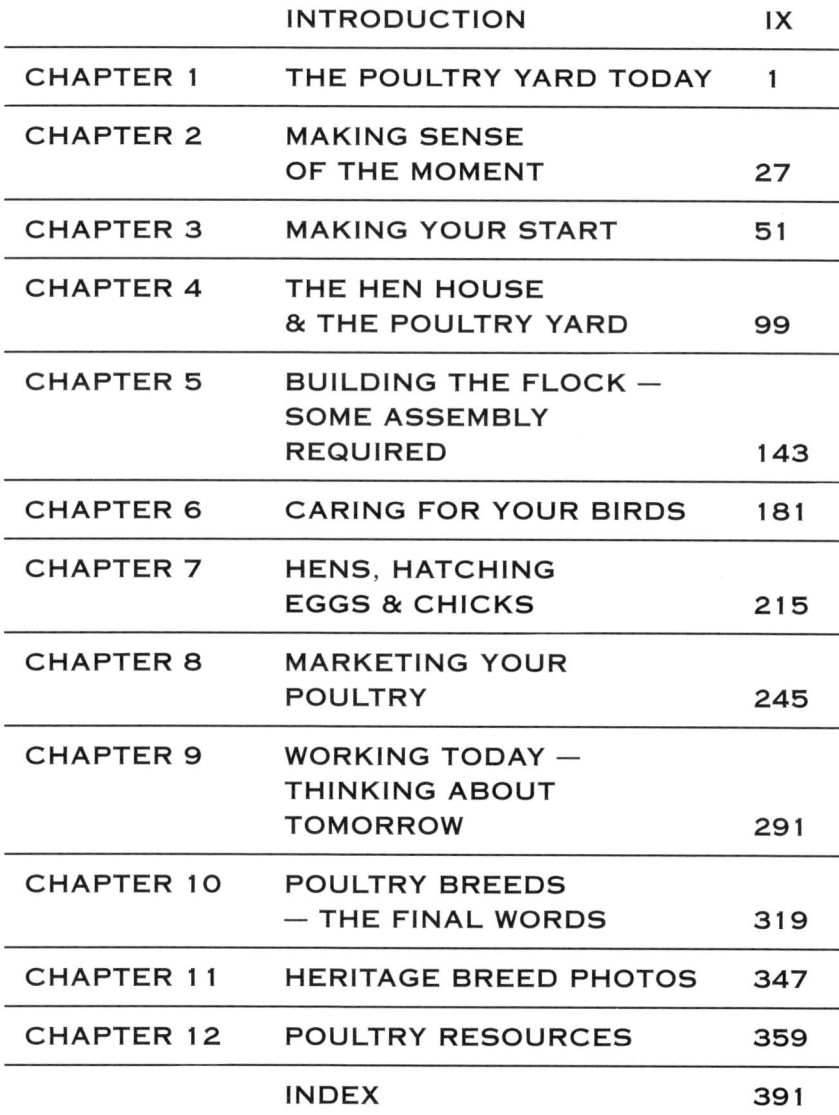

	INTRODUCTION	IX
CHAPTER 1	THE POULTRY YARD TODAY	1
CHAPTER 2	MAKING SENSE OF THE MOMENT	27
CHAPTER 3	MAKING YOUR START	51
CHAPTER 4	THE HEN HOUSE & THE POULTRY YARD	99
CHAPTER 5	BUILDING THE FLOCK — SOME ASSEMBLY REQUIRED	143
CHAPTER 6	CARING FOR YOUR BIRDS	181
CHAPTER 7	HENS, HATCHING EGGS & CHICKS	215
CHAPTER 8	MARKETING YOUR POULTRY	245
CHAPTER 9	WORKING TODAY — THINKING ABOUT TOMORROW	291
CHAPTER 10	POULTRY BREEDS — THE FINAL WORDS	319
CHAPTER 11	HERITAGE BREED PHOTOS	347
CHAPTER 12	POULTRY RESOURCES	359
	INDEX	391

• INTRODUCTION •

FEW ARE THE TIMES IN MY LIFE when we haven't had at least a few chickens around our farm and in recent years they have grown to be one of our major pursuits. Small flock poultry production was very much a part of the farming mix when I was a boy and now the purebred poultry flock has returned as a viable enterprise for the well-diversified, modern family farm.

Egg and poultry meat production were the first livestock ventures taken up by the vertical integrators, but the nation's rich pool of chicken genetics has endured despite the corporate sector's best efforts to reduce the poultry population to a handful of white feathered hybrids producing medium to a few large white eggs and irregular chunks of chicken meat.

There are more chickens than any other livestock species on Earth and more people keep chickens than any other livestock species. Here and around the world, chickens are kept in near infinite variety. But among the most enduring and productive are the group of heirloom or heritage purebreds with long ties to the small, family farms of the United States. They are the birds that spawned the fabled butter and egg trade of the 19th and early-20th centuries.

Breeds like the Rhode Island Red, the Dominique, and the Barred Plymouth Rock are near legendary and in reclaiming and revitalizing these and other heritage breeds, family farmers will do much to regain their historic role as the poultry producers of first choice and greatest respect. Unfortunately, many of these breeds have endured for a half-century or more with little or nothing in the way of selective breeding for greater productivity and life on range or in the larger poultry yard.

With a purebred flock of one of the heritage chicken breeds a farmer/breeder can create a fully sustainable poultry venture that can be bred on and upgraded for generations. These breeds and their output are the possessions and provenance of the American farmer and in preserving this genetic resource and returning it to the productive role for which it was bred, the position and income potential of the farmer/breeder will also be strengthened.

The birds of our grandparents' poultry yards can and should be put to use producing for today's marketplace that so values locally bred and grown, historically significant agricultural production. There is new life in the old poultry yard now and much of that is due to those great old farmyard birds and returning them to their earlier promise.

— Kelly Klober
Silex, Missouri
May, 2010

• CHAPTER 1 •

THE POULTRY YARD TODAY

PARTRIDGE
• WYANDOTTE MALE •

PARTRIDGE
• WYANDOTTE FEMALE •

WHITE PLYMOUTH
• ROCK MALE •

WHITE PLYMOUTH
• ROCK FEMALE •

WE ARE ABOUT A DECADE into what some have called a chicken renaissance. A few years back range broilers and loose-housed laying hens emerged at a time when consumer concern about food sources was great and family farmers were looking to reclaim something of their heritage as diversified and artisanal producers. The range broiler was little different from your grandma's frying chicken, but it soon became the foodies' chicken of choice. Since then the range broiler has continued to evolve from humane and naturally produced bird to ever more exotic forms including "milk-fed" and even "baby" chicken forms that grace restaurants where a plate of chicken can set you back three figures. It would delight my grandparents to know that their skimmed milk-fed chickens of yore are now gourmet fare.

The egg of the moment is some shade of brown, cage-free, additive-free, possibly organic and is marketed now both for eye appeal and social correctness. Historically, different regions of the country have expressed a preference for eggs that were either white-shelled or brown-shelled. Boston was a white egg town and New York City a brown egg town back in the days of the butter and egg deliverymen. With the advent of vertical integration and colony, caged laying houses, nearly all commercially available eggs became white-shelled. The egg "industry" chose layer lines based almost entirely on the White Leghorn breed because of their high egg productivity per bird. They then bred them down in size through selective breeding to create laying lines that could be packed quite tightly into all-wire laying cages. Their lighter weight also caused them less foot damage from life spent entirely upon a wire mesh floor.

With the White Leghorn you get white eggs and in such large numbers that they are now largely seen as a loss leader item in the retail food trade. The brown egg became something of a rarity and more and more they were becoming viewed as a product of true mom and pop, grassroots agriculture. The perception was that brown and non-white eggs had to be a better egg because you had to go out of your way to find one. Well, this is partly true. A number of years ago there was an early surge of interest in Ameraucanas, the Easter egg chickens. It was thought that their mostly green-shelled (and a few blue) eggs were healthier and far lower in cholesterol than the brown or white eggs on the market.

When university testing was finally done to investigate this notion — some real surprises were revealed. It was determined that eggshell color had absolutely nothing to do with the egg's nutritional content. Regardless of shell color — white, parchment, brown, dark brown, blue, green or pink tinted — they all tested the same.

The real surprise that did emerge when testing of the different eggs was completed was that all of them had lower levels of cholesterol than had previously been believed. Earlier test results had been wrong. Within reason, eggs can be a part of nearly everyone's diet and a full eighty percent of the population need pay little or no concern at all to eggs as a source of harmful cholesterol.

Still, the colored egg continues to be perceived as something unique and better. When marketed with an emphasis on other factors such as being raised with cage-free housing or additive-free rations, the colored egg has found a niche as a truly premium product. Prices have been found that approach four dollars a dozen for organic brown eggs and nearly $2.50 a pound for (four-pound weight) range broilers certainly have fueled some of the interest in this new era of poultry keeping. In addition to humane poultry treatment and healthy eating, there has also been a growing concern and interest in rare and heirloom poultry breeds.

Many of the rare, heirloom or heritage breeds have been around for decades in North America and others have been brought in from abroad in more recent times. But these minor breeds were found only in small, rapidly declining populations. A number of historic breeds have gone totally extinct over time and many others exist only in populations of a very few hundred or less.

Why the recent interest in heritage poultry breeds?

Interest in these small, rare breeds began to grow a bit over a decade ago as a result of several factors.

From a meteoric high, interest in exotic animals generally was winding down due to increasing regulations, increasing costs to maintain, and growing numbers that were driving down both the demand and price for breeding animals. The early market for popular breeding animals is always high and volatile and then levels off.

Rare fowl breeds caught some of the same energy as the exotic animals had, but also had the benefit of being far simpler to acquire and maintain. They sold well and briskly, had an even broader market base than the exotic animals, had more practical uses, and far fewer regulations and issues associated with their upkeep.

The birds are not just economically, but historically important. The first round of interest in rare poultry "exotics," in the early nineteenth century, basically launched the entire modern livestock industry. The importation of poultry breeds like the Shanghai (today called Cochins) launched a huge tide of interest in poultry keeping rather than mere chicken owning. In succession came the development of breeds like the Rhode Island Reds and White Plymouth Rocks and the emergence of the broiler industry based on such early breeds as, first the Java and then the Black Jersey Giant.

This interest in poultry propagation led to the creation of the first major livestock organization in the United States, the American Poultry Association. The first livestock exhibitions, comparable to today's state fairs and type conferences, were designed to display and promote the keeping of well-bred, purebred poultry.

Today, one of the fastest growing and most popular youth livestock projects is the breeding and exhibiting of broiler and purebred poultry.

Interest in the rare and heritage birds has emerged at the same time as growing interest in simplifying and gaining greater control over one's life and home. The brown egg from the farmers' market, range broilers from a nearby small farm, and even the backyard chicken flock all handily fit in with these desires for a more healthy and sustainable lifestyle.

Poultry interest has often been recharged with the appearance of new and different breeds (Marans, for example) and developments such as the arrival of very dark brown-shelled eggs. These and other new introductions into the poultry market continue to spark more interest.

Minor and rare breeds interest continues to grow because the birds have quite modest space needs and rather low start-up costs. They also offer a rapid turn-around on investment, sometimes in a matter of just weeks.

Just what are the rare and heritage or heirloom breeds? Essentially, they are the breeds and varieties that were left in the wake of the industrialization of mass-produced poultry that began about seventy years ago.

At that time, the commercial sector became enamored with all things white. The white egg-laying strains derived almost entirely from the White Leghorn breed, and white feathered and yellow skinned broilers derived from line crosses of the White Plymouth Rock and White Cornish breeds. Such birds dressed cleanly and met the American preference for yellow legged and skinned table fowl.

A slim few other breeds did persist in fair numbers and continued to be offered by a relative handful of hatcheries that still remain and ship nationally. Breeds most commonly offered were Barred and White Rocks, Buff Orpingtons, Silver Laced Wyandottes, Brown (Light Brown, actually) Leghorns, New Hampshire Reds, and Rhode Island Reds of somewhat dubious breeding and referred to variously as "performance" or "production" Reds. Their so-called "rare" breeds would be Brahmas, Ameraucanas, a few Polish varieties, and a handful of bantams. These were what have come to be known in some circles now as "hatchery," "industrial" or "commercial" bred poultry varieties.

The genetic integrity of these varieties was brought into question all too often by bracketed quotations in the poultry catalog descriptions noting that birds of certain varieties are labeled "not for exhibition" or "not for project use." Most hatcheries no longer kept breeding birds in any number and a number of hatcheries still may buy their eggs for certain breeds largely as needed and from but a single breeding flock. As a result, a hatchery in Minnesota and one in Texas might sell you birds all of the same breeding.

The bird fancier or exhibitor sector have kept alive a great many different breeds and varieties, although their primary focus was not egg or meat production but to breed for extreme, showroom type. Their selective breeding was largely for extreme size and visually intense feather quality and color. This exhibition breeding came often at the expense of the performance traits for which these breeds were initially bred. It is true, that even those breeds that looked like they had strolled forth from the pages of a Dr. Seuss book were initially developed with mostly practical intentions. Elaborate head crests, for instance, were bred for life in cold climates. The crests were meant to give added protection to their combs and delicate head areas.

These bird fancier men and women did a great service in keeping these breeds alive in viable numbers. They still form a large network of dedicated breeders and have fostered an infrastructure that has been vital to the resurgence of the heritage breeds. Still another group of enthusiasts are preserving the purebred poultry breeds for that old time stockman's virtuous calling, "the love of the breed." These enthusiasts hung on even though they were being both denied and ignored. Now it is often their flocks that are being rediscovered by modern preservationists.

The American Livestock Breeds Conservancy (ALBC) defines a rare breed as one that exists nationally in known breeding numbers of fewer than five hundred birds. Their list includes birds that were once quite popular such as the Ancona and some such as the Russian Orloff that never existed here in great numbers.

"Rare breeds" is also the category that includes recent arrivals such as the Penedesenca, rare color varieties of certain breeds, and breeds such as the Single Combed Rhode Island White that never really had wide acceptance or even official American Poultry Association sanction. Due to their fecund nature, rare breeds don't stay rare very long if they catch on with enough breed enthusiasts. However, sometimes in the swift spiral up in popularity, too many of the wrong kind will be kept as breeding birds. With the growing consumer popularity for the very deep brown table egg, breeds like the Maran have begun to grow greatly in both numbers and color varieties. I recently saw new additions of both birchen and silver varieties displayed. Another dark brown egg layer, the Barnevelder, has on the other hand remained somewhat static in its numbers because there have been breeding problems within this breed. The double-laced color pattern is a quite complex one and they are starting to build from a truly quite small population. As a result, some have said that if you need one Barnevelder rooster you'd better buy five.

The heritage breeds are the birds of our historical small farmers both here and abroad. Some of these breeds are truly quite venerable. The Dominique is believed to have its origins in the seventeenth century and the Black Java is believed to be the oldest or second oldest American breed. Some have an almost romantic past such as the Dorking and the Pit Game that have been kept both by Presidents and rogues. Others have a long history as a truly farmers' fowl such as the Wyandotte, Minorca, and even the sprightly Hamburg.

The standard sized Leghorn has been bred in over twenty different colors and patterns and the Wyandotte in over a dozen colors. The white and black patterned Exchequer Leghorn, the Leghorn breed bred in Scotland in the early 1900s, may be one of the rarest of breeds on the scene right now. We breed blue Wyandottes. Some colors of both Wyandottes and Exchequer Leghorn breeds have not been seen in many years, but the chance exists that populations of them could still be found at any time in some remote corner of farm country.

Just a few years ago an elderly gentleman in the state of Mississippi contacted the Society for the Preservation of Poultry Antiquities for help

EDUCATION ON THE FARM

Having grown up around the farming industry in Colorado, Karen Weppner of Promised Land Family Farm always knew that she wanted to someday have a farm of her own. When Karen was a child her father had a small farm where he raised and sold Clydesdale horses. Though it wasn't his main source of income, just being in the farming atmosphere and around animals sparked Karen's affection for farming. So after years of dreaming, Karen and her family moved onto farmland in Coeur d'Alene, Idaho five years ago.

Like her father's Clydesdale business, Karen's farm is not her main source of income; instead her farm serves as an outlet for education in her community. Because Karen does not devote full-time to the farm, her mindset when it comes to her animals is that they ideally provide their own care and need little assistance from Karen and her husband.

When Karen started up Promised Land Family Farm, it was the only one of its kind in her area. With her town lacking a zoo or any other natural wildlife center for visitation, people began visiting her farm and learning things about animals and planting that they never would have learned otherwise. Also, in her desire to educate, Karen began informing her visitors of the negativities that come along with processed foods that consumers buy from grocery stores every day. Farm visitors would cringe when Karen informed them that almost 99% of the turkeys bred for their Thanksgiving dinners were the same breed and that those turkeys can't even reproduce without human intervention. Karen says that people know very little about the origin of the food they eat and that her farm is trying to educate in that regard.

Because of Promised Land Family Farm's locale, Karen's land is not the best for large-scale farming. Instead, Karen's motto is "a little bit of everything." Her farm houses not only heritage and heirloom breed poultry but also sheep, llamas, horses, pigs and goats for the neighboring townspeople to come see. Located a short five miles out of town, it's easy for people to make their way to Karen's property and experience all that a real-life sustainable farm has to offer. And more than anything else, Karen's real desire is to inform and educate others of the benefits of local food consumption and farming.

When it comes to deciding which animals to breed and which animals to stay away from, Karen has a relatively basic standard. If an animal is able to breed and survive the weather conditions on her Idaho land, then they get to stay. In terms of her breeding techniques, Karen likes to stick with nature and limits the use of artificial incubators. And if a mother hen will not brood her own eggs then Karen can typically get one of her Silkies to raise the eggs. And this doesn't just go for chickens; Karen says that her Silkies will raise almost anything in an egg, including geese and even turkeys. Because Karen does little to alter the natural breeding and raising of her animals, there's a wide variety of heirloom and heritage breed poultry on Promised Land Family Farm including Ameraucanas, Plymouth Barred Rocks, Speckled Sussex, Silkies, Bourbon Red turkeys, Indian Runner ducks, and Tufted Roman geese to name a few.

In the five years that Karen has been on Promised Land Family Farm, she's had a lot to get excited about. When her farm was able to have a natural bonding, breeding, and brooding of Bourbon Red turkeys, (on the "Watch" list of the ALBC) Karen said she was completely overjoyed. But more than anything else, Karen says that her greatest moments are when families that visit her farm turn around and create farms of their own. She says there's nothing better than knowing that she had a part in bettering the community in this way and that that's the true driving force behind Promised Land Family Farm.

In the relatively short time that Karen has had her farm, she says that her challenges have been many. Her biggest challenges are, oddly enough, the most natural – seasons. She says that when you're just a consumer you don't realize the effect that weather and climate have on food production. When the "Eat Local" movement first kicked off Karen said she had people calling for vegetables in January to which she had to reply, "Try us in July." And that's simply the reality of farming.

When it comes to advice, Karen has a lot of it. What she stresses most is that if you're looking to get into this business, then the best thing to do is to get out there and see for yourself. And if there's not a local sustainable farm to visit, then read! There are numerous books focused on heritage farming and also countless pages of information on the Internet. But above all else

Karen says, "Experience is your best teacher." Eventually you can only read and observe for so long until it's time for you to get your feet on the ground and begin. Of course there will be downfalls and it will be tough, but it's not impossible. And whether you're a farmer or just a consumer looking to make a change, you can learn one important thing from the "classrooms" of Promised Land Family Farm — that everyone can make a difference.

Promised Land Family Farm can be found at www.promisedlandfamilyfarm.com *or by phoning* *1-208-512-3305.*

with dispersing his many small flocks of different chicken breeds. Among his breeds were a handful of White Houdans, a breed many thought could no longer be found in this country. From a flock of fewer then a dozen birds, a major breed restoration project was started by a group of cooperating breeders. Such things happen often enough to keep a real fire burning in breed preservation circles.

The Chantecler, the breed developed very early in the twentieth century for productive life on the family farms of Canada, was likewise once believed to be extinct. Efforts were actually underway to recreate the breed from its founder's notes when small populations of Chanteclers were discovered to still be in existence. Beyond the original white variety it is now also being bred in buff and partridge color patterns.

Some breeds just never caught on in a big way or were bred to closely resembling other, more popular breeds. This is reflective of a question often raised about the preservation efforts to save "so many" breeds. Why save them all? Are they all really needed?

I have long been partial to the white breeds of chicken, but to some people they are all just so many varieties of plain vanilla. In the late 1920s much effort was given to launching the Rhode Island White as an egg layer option for farmers with small- and medium-sized laying flocks.

These Whites were and are still very productive birds. In fact, the most productive birds we have ever owned were a strain of Single Combed Rhode Island Whites acquired from Iowa.

The white-feathered, brown egg-laying category was already a quite populous one when they came along. Efforts by the Rhode Island Whites to emerge and then stand apart from others in this category faced truly strong competition from such firmly entrenched and established breeds as the White Plymouth Rock and the White Wyandotte among others. The breed politics of the day and the economy went against them and only the Rosecombed variety of the Rhode Island White went on to be sanctioned by the American Poultry Association. The Single Combed Rhode Island White is still at best a living historical footnote. A poultry footnote existing in quite small numbers, but of such quality that it has endured well out of the spotlight for nearly three-quarters of a century.

Another white bird that fell victim to low numbers and competition from earlier, established populations of comparable breeds was the Lamona. Until a recently reported rediscovery, this breed was actually believed to be extinct. Some creative people had researched its origin and were actually contemplating recreating this laying breed.

Do we need all of these white breeds and all of the colored ones, too? Probably not, nor do we need to make the very foolish mistake of thinking we can float a national poultry industry — especially in a nation as broad and environmentally diverse as this one — with just three breeds of chicken.

A few years ago it was postulated that if America continued with an ever-narrowing gene pool it was ripe for a genetically based catastrophe of epic proportions. What would happen if a long hidden genetic fault emerged, a new disease blew in with a hurricane or circumstances of production suddenly shifted due to economic or environmental changes? The poultry industry lacks the bred in flexibility to change. It has been held that only after a frantic search of Third World countries might it be possible to produce the genetic material needed for a major rebuilding of the nation's poultry flock.

Virtually every poultry breed, no matter how closely it resembles another, was developed for very specific ends and purposes. The French strain of the Cuckoo Maran was developed through selective breeding for life in a lowland, damp environment. Breeds like the Sussex and Orpington were developed for specific markets and often took their names from nearby large cities. Some were bred for egg production, some for growth and meat-types, some for vigor and an easily sustainable size in a harsh environment, some for feathering with which to better cope with a local climate, and many for a combination of purposes. Like all other human creations, a certain amount of consideration was given to breeding aesthetic appeal at some level. When allowed to revert back to the wild type, which is something they can do in as few as four generations, most birds will develop the Black-Breasted Red color pattern of their jungle fowl ancestors in Southeast Asia.

The date when chickens first became domesticated is still a topic that is much debated. Before poultry domestication it is safe to say that humans robbed the nests of jungle fowl, caught them in snares, and plucked them from roosting places after dark.

It could be said that some of the very earliest livestock pure breeding done in this country, and even earlier in Europe, was to produce game fowl for the fighting pits. It is not for us to discuss the wrong or right of this pursuit here, but the role of the Pit Game in the history of poultry keeping has to be acknowledged. The historical importance of the Pit Game was noted in Mr. Alex Haleys' landmark book, *Roots: The Saga of an American Family*. Certainly, Presidents in this country and aristocrats abroad kept fighting birds. The lessons learned in breeding these fighting birds were certainly applied to other types of poultry and livestock.

Some Gamecock lines still around today were known three hundred years ago and were brought here from countries where the practice had been known for centuries. As a group the Pit Games now present a truly great conundrum to those involved in poultry preservation work.

No breed is more deserving of attention than any other and while the practice of pit fighting is forbidden in many areas of the United States, what will eventually happen to the pure and long-enduring gene pool these birds represent? No birds are more vigorous than the Gamecock birds, more historically important, naturally broody, bred in a great variety of types and colors, and are fair layers of what many describe as the best tasting of all eggs.

More than one rare breed population has been bolstered by a one-time infusion of Gamecock breeding from a Black-Breasted Red patterned hen. Back breeding will bring a line back to desired type with a little dash of Gamecock keeping the "blood" warm and strong.

Those really sincere about poultry breed preservation work will have to consider this group of birds. It is not illegal to own and breed them. Many are truly quite beautiful and while there needs to be some respect

SELECTING A POULTRY BREED FOR YOUR FARM

Walk among the breeding pens on our farm and you will see a chicken breed or two commonly referred to as "dual purpose varieties." I have always felt about this usage of the term dual-purpose much the same way my banker feels about the term "the check is in the mail."

Yes, all chickens lay eggs, and sooner or later, nearly all growing chickens get big enough to fry. To expect one breed or line to do both of these well, however, may be asking just a tad more than Mother Nature can or will easily bring forth from an egg and wrap in feathers.

Begin breeding for improved egg production, and noticeable changes in bird type and size will follow. As laying improved in our one line of Wyandottes, the hens became a bit smaller, and there were some subtle shifts in frame size and outline. To improve growth and meat yield you will likewise select for different factors, including frame size, growth rate and muscle yield. Cornish and Giants do not lay like Leghorns, nor should they be expected to do so. With selective breeding, virtually any trait can be made better within a breed or line, but just as they can't design a vehicle suitable for both moving couches and racing at Indy, you are not going to find birds that can fry at six weeks and begin laying 300 eggs a year at 20 weeks of age.

A decade and more in the field of public works has taught me that any job is done best only when done with the right tool. If you want eggs, select breeds and lines bred for egg production. If you want poultry meat, select breeds and lines bred for growth and yield. If you have a strong market for both, then you should seriously consider maintaining two separate flocks. It would be inefficient and less than optimum production if the small farmer were to do otherwise.

Yes, I know I am incurring the wrath of supporters of some breeds and trouncing on the idyllic beliefs of others. Still, a businesslike approach must be brought to poultry keeping, just as to any other venture on the farm.

With a purebred flock of chickens you can reasonably expect to produce six cockerels for every four pullets hatched. Depending upon breed, young males will reach a handy, dressable weight at somewhere between 12 and 16 weeks of age. They won't outdo Cornish X broilers, but then the bird to do that will have to be sewn together on Dr. Frankenstein's lab table. These purebreds should grow into some quite nice roasters and may present the alternative for those wanting a change from turkey. My grandmother's choice for frying chickens was young Leghorn roosters harvested from the spring crop as summer's warm days grew to an end. They didn't yield by the "bucketful," but they still tasted mighty good.

A two-flock farm need not have birds by the hundreds parceled out from here to breakfast, either. Actually, a meatbreed flock can produce hundreds of chicks from a relative handful of breeding birds. A one-rooster flock should be more than adequate for most producers of a thousand or less range broilers each year. I say one rooster, but be sure to winter enough birds to insure a viable breeding flock for the next year. You don't want even a small flock of hens standing around because your only rooster gave up the ghost.

Many in the chick-production field will sell their breeding flocks at the end of each hatching season (generally May or June) and hold back young of the year for next season's breeding flocks. It is a practice that certainly reduces maintenance costs and winter chores.

Still, I would be very reluctant to let go of birds of exceptional merit. Being too exacting in paring numbers can sometimes result in future losses and setbacks, as well. I have a friend who put substantial time and effort into assembling a flock of rare White Langshans. All was going well until he lost all four of his breeding males in the space of a single week (such losses can be caused by illness, storm damage, predation and other factors). His flock then stood unproductive for many weeks while he scrambled to find even one breeding-age male of good quality.

Pure breeds with real potential as meat birds would include White Rocks, Delawares, New Hampshire Reds, White Giants, Standard Cornish, Buckeyes and White Wyandottes. Continental breeds such as the Sussex, Dorking and Orpington have long histories as table fowl abroad, although they lack the yellow skin and feet so favored by American consumers. There are also many other colors in which Rocks and Wyandottes are bred. Another friend of mine is beginning a Giant breeding program to develop a meat-breed strain with size that will go head-to-head with the domestic turkey.

Your choice of meat breed can then be paired with one of the egg laying breeds to give your family farm a true one-two punch when it comes to poultry production. Laying breeds run a wide gamut, from the legendary production of the Leghorn and Ancona to the rarest of the rare such as the Welsummer and La Fleche. Those interested in breed preservation work might wish to consider varieties like the Exchequer or Red Leghorn.

Fulfilling two economic needs with one bird would truly be an ideal, but not a very realistic one. Dual-purpose breeds have been touted in many different livestock species, but at best they have been compromises — and they could and did compromise farm production in far too many instances.

It appears that fast approaching is a new era in family farm poultry keeping. The heritage breeds are being returned to their traditional role of production fowl working in range and open-housed environments. It is a role not at odds with exhibition breeding, but it will have many producers more strongly emphasizing those factors most associated with improved economic performance.

Hybridization to increase performance was the tool of the vertigal integrators, but it was made so complicated in its structure as to be carried out only by the largest and most complex of corporate producers. They thus have a most solid lock on that course of production. With purebreds, the independent producers retain the greatest control possible over their flocks, their farms and their future direction. Through selective breeding, simple applied genetics, they can exercise a great level of control and produce birds to the most exacting needs of their home farms and local, directly served markets.

> They can use the variability to be found in the remaining pure breeds to create working flocks that will produce eggs, or broilers and roasters, in a most efficient manner. It is made efficient by selecting the breeds most suited for the production needs at hand.

accorded to their gritty nature, they are not hard to keep. You won't pen them with your Buff Orpingtons, but they are really fairly tractable birds.

Good preservation work will eventually eliminate a breed's rare and endangered status. Some of what has held these breeds back is the thinking that they are the bailiwick of just a certain few breeders. Not true! Nearly all were developed for real world tasks in farmyards around the globe. They were bred to produce eggs or poultry meat, self-replicate, forage for at least a part of their existence, and to fit comfortably into a mix of cropping and livestock ventures that then went on to create a successful and well diversified family farm.

Initially they were developed with at least a dash of breeder or producer ego in the mix, but they were nearly all meant to be working fowl. They deserve then to be taken down off the shelf not just to keep their genes alive, but also to continue breeding them up for improved performance and type. Those that own them need to have a plan far beyond merely keeping a handful of them alive as a museum curiosity item.

Rare breeds also don't deserve to be held to the same production standards as the majority breeds. You won't get the Polish or the rarer Rock varieties to lay in the same league as today's production bred Leghorn strains. Although in time, these rare breeds can possibly be bred to be 180 to 200 egg producers per hen per year. There are some good true Rhode Island Reds now at around 160 to 180 eggs per hen per year. As for meat birds, nothing so much epitomizes "Frankenfood" as today's Cornish X broiler. There were and are alternatives in the form of such pure breeds as the Rocks, Delawares, and Wyandottes. Nearly all of them in this group have not been selectively bred for growth and meat yield for well over half a century.

There is a term that I think is very unfair and misleading and is used now to describe some breeds that serve more than one use. If you buy into the "dual-purpose" chicken argument you probably have a whole closet full of one-size-fits-all clothing, too. There are breeds that lay a moderate number of eggs and do a fair job of growing to size, but you won't ever find a Leghorn hen in a White Cornish body.

The two roles for the chicken are meat or egg production and attempts to combine them will compromise both. If your greatest need is for eggs, get started with a breed known for this trait and then continue to upgrade it with selective breeding. For meat production select a larger framed, heavier meat variety.

A good layer-type isn't necessarily a good meat-type. All chickens lay eggs and all chickens will eventually fry. It's just that for some breeds ninety eggs is a good year and for others a bit under four pounds is a mature hen weight – far from the maximum possible for poultry.

We began in the heirloom field with a set of white Wyandotte chicks. The breed has a reputation as a fair layer, it produces the always-popular brown egg, and its feathering and rosecomb were conducive to our simple, unheated housing facilities. As time passed and we selected evermore for improved egg-laying performance, the birds grew a pound or two smaller and a bit more vigorous in their type. They never did lay like Leghorns, but they did come to lay better than their more immediate ancestors. In a similar vein, birds bred for the current trend of larger sizes for the showroom have seen some reduction in egg output — even in breeds long associated with egg production.

Preservation work is at the point where most of the breeds have at least been taken off of the shelf, but the breeders involved are still just blowing the dust off a great many of them. Some are from very narrow gene pools, some populations are terribly inbred, some have real type issues, some are still held in a very few hands, and a few are still more myth than substance. They are a hot property if for no other reason than that old-time economic concept of supply and demand.

These rare and heirloom breeds are about as high profile as anything to come down the livestock pike in a decade or more and they have it in them to be far more than simple, feathered fads. These birds were once the heart and focus of a very big and lucrative business — and still can be once again.

Poultry keeping spawned the thought behind and many of the processes now at work in all types of modern livestock production. The check from selling eggs stabilized the family farm for decades. Laura Ingalls Wilder even financed her early writing career with income from her work on her and Almanzos' modest Missouri poultry farm.

The egg check may be about as close as the American family farmer has ever come to having a weekly paycheck. At one time rural children would even be given an egg to trade for candy or geegaws during the family's trip to town to market their production. At one time, and it continued well into the twentieth century, the worth of a good breeding bird or clutch of fifteen hatching eggs was equated to a day's wages for a working man.

Until the post–World War II years, a big poultry flock was three hundred to five hundred hens, and chickens were an everyday fixture on many farms from Maine to Malibu. Major poultry centers such as the Del-Mar Peninsula on the east coast and the area around Petaluma, California were coming together and starting to drive production, but it was still largely chicken production on a human scale.

During my youth in the early '60s this smaller focus was in decline, but traces of it were still to be found here in the Midwest and elsewhere. I can still recall the old Alderson Brothers' mercantile that took up nearly all of one side of Main Street in our little town. They sold dry goods, groceries, livestock feeds, and they bought and sold chickens and eggs. From them and our neighbors my grandparents bought six to ten thirty-dozen cases of eggs (all white-shelled even then) that they sold each week along their egg route in St. Louis County, Missouri.

All along the western wall of the Alderson's red brick building was a loading dock and a series of chicken pens. On some days the pens would be filled with literally hundreds of spent hens, roosters, and fryers, when in season. I spent many youthful hours there just walking the pens looking for birds of rare color or breeding.

Gradually, the birds there grew fewer and fewer in number and then the pens ended up standing empty for weeks at a time. Finally, the day came when the old pens came down and one of the major landmarks of my youth just slid away. Today there is no place that I know of in the state of Missouri where an independent producer can set a case of eggs or coop of chickens on a loading dock and receive a check.

Within a short distance of our home and well into the '60s we had a neighbor with a hundred or so Ancona laying hens, another with three to four hundred Austra-Whites (Black Australorp male x White Leghorn female). Also near home were two sisters and their brother and earned money by the sale of eggs from their flock of White Leghorns. A more distant neighbor had flocks of Reds and Rocks. Nearby was even one of the last free ranging turkey flocks in the state. Back then the birds weren't yet shut away in a sheet metal gulag like they are today.

Many of today's rare breeds were still the basic inventory of a great number of hatcheries back then. Most of those hatcheries are gone now — from the Steele family hatchery that anchored one corner of the main drag in the next town over from ours to truly national firms like the Berry Hatchery of Clarinda, Iowa. Both of these and many more are gone now. The Berry catalog was the first piece of mail I can ever remember receiving. I must have read it a thousand times over. Containing the full gamut of poultry from "Danish" Brown Leghorns to breeds like the Lamona, which has come dangerously close to leaving the pale of this earth, the catalog was a storybook of what poultry farming was and perhaps could be again.

Around this earth of ours more people now keep chickens than any other livestock species and they exist in greater numbers than any other species. Spawned in large measure by the revived interest in heritage or heirloom breeds, the rare and emerging breeds have seen a steadily documented yearly growth of three to five percent in the number of birds being kept outside of confinement production.

The interest in these breeds is great — a marvelous boost and flourish of a valuable resource, but at the center of this interest is something more. It is a full circle return to a more humane and artisanal approach to production agriculture. It isn't mere production of a commodity, but a revival of the creative process that was once a very successful, small-scale agriculture. Back then you didn't succeed by producing more of something, but rather by making that something better. In my grandparents' time, big and good were not synonymous and we're starting to see that they are not now, either.

In the fifty years or so from the teens to the mid-sixties of the twentieth century, there were tens of thousands of small flock poultry producers and breeders. The latter kept purebred breeding flocks the way pedigreed dogs and blooded horses are kept now. Names like Sturgeon and Thomford were associated with breeds like the Barred Rock and the Rhode Island Red the way the fabled King Ranch is now associated with the production of registered cattle and Quarter horses.

Such an age of poultry keeping can and should come again and it just may grow out of this present-day interest in the rare and heritage breeds.

What must not happen is an ongoing effort to simply build on their rare and unusual aspects. Such novelty markets are nearly always short-lived and too often leave a great many with a bitter taste.

What no one wants to see happen to this interest in heirloom poultry is for it to become the next Vietnamese pot-bellied pig fad. There are certainly seeds of the pot-bellied pig scenario within the heirloom poultry world with a great many people now busily jumping from one "hot" breed or variety to another. If you have to have some of these birds simply because they've been pronounced rare or because Martha Stewart has some or because of recent selling prices then you don't need them and they really don't need you!

The goal for rare poultry should be to make them "un-rare." To that end it should be about restoring them to the roles from which they were originally bred and refined. Many should work diligently to preserve in these rare breeds the vigorous nature and durability that has enabled them to survive to this point, upgrade them through selective breeding, and then to get them into as many new hands as possible.

SEASONAL CONSIDERATIONS FOR POULTRY BREEDING

The production year for a producer with purebred poultry may begin as early as December of the preceding year. Mid-to-late winter is when breeding birds are sorted and selected and breeding pens put together.

Some will match males and females based on the time-honored practice of breeding genetic strength to genetic strength — a good layer to the son of a good layer, for example. Others selectively breed for a desired strength, breeding a bird strong in one area to birds somewhat lacking in that trait. Many will seek some added simplicity in this task by following a "rolling mating plan." They will breed the best pullets of the year to the best males of the preceding year and the best cockerels of the year to the best hens of the preceding year.

Most baby chicks are hatched in the months of February through May. Eggs pure for the mating can be gathered 14 days after the desired male has been introduced into the breeding pen. A female can produce fertile eggs by a male for several days after he has been removed from the breeding pen. This is certainly something to remember when rotating males or restructuring matings.

Common practice was and is to produce chicks that will begin laying in the fall of the year in which they are hatched. Chicks hatched late in the year will grow and develop more slowly due to factors such as temperature stress, shorter hours of daylight and changing seasons. With the rare breeds we often hatch late into the year to build numbers and satisfy demand, but have encountered more challenges with the late hatches. Some will sit out the heat of late summer and come back with a few fall hatches.

Old rules of thumb hold that you should start anywhere from three to five chicks for every bird you eventually want to add to or replace in the breeding flock. If buying sexed chicks these numbers might be halved, but as time passes and numbers are built, the culling process should become ever more severe and demanding.

Many are now finding strong markets for started chicks of several different ages. Started chicks can be fairly safely sexed by sight, and as early as two weeks of age with some breeds. Many producers have found very strong markets for small lots of ready-to-lay pullets or soon-to-be-ready ones, birds in the range of 14 to 20 weeks of age.

These birds represent a substantial investment in time, facilities and feed, but they are truly a premium product of the family farm. All of these factors are reflected in their selling price — at local markets I have seen started pullets offered for sale for $5.50 to $8.50 each. Every time I see started pullets offered for sale I am reminded of an old Missouri cattle auctioneer who greeted every set of heifers into his sale ring with, "Lookit here, boys, they have their whole lives ahead of them!" Such is the value of a started pullet, all of the hard work of getting her started has already been done, and her entire egg laying career lies ahead.

The niches before poultry producers today are many and varied. To fill them, the producer must seek out the best birds for the task, breed them to a high standard, and spin a fair amount of business acumen into the mix.

Originally published in Acres U.S.A.

CHAPTER 2

MAKING SENSE
OF THE MOMENT

HOUDAN
• MALE •

HOUDAN
• FEMALE •

WHITE CRESTED
• POLISH MALE •

WHITE CRESTED
• POLISH FEMALE •

POULTRY KEEPING IS GETTING A LOT OF PRESS lately with numerous articles in farm and rural lifestyle magazines and newspapers. Also at many farming-related meetings and seminars, the best-attended sessions have been those with poultry themes.

Chickens are at a level of popularity many thought would never come again. These are truly exciting times for poultry raisers, but remember that old Chinese curse, "May you live in interesting times!" The chicken's relationship with the Internet is rather like a new age version of that old riddle about the chicken and the egg and which came first. Now there are online auctions for hatching eggs, baby chickens, and adult birds. Chat rooms and online forums are rife with poultry topics and all sorts of facts, myths, fads and rumors have been spread out from there.

There is now even an international trade of sorts in poultry ideas and genetics. New hatcheries have appeared, new poultry-themed magazines are being published, poultry groups are growing in membership, and support infrastructure is again forming around these smaller-scale producers. Now in many farm supply catalogs, more pages are given over to supplies for poultry than for hogs or sheep.

Feed and farm supply stores now even have baby chick days from early spring through late fall. A very brisk trade has grown up around the rare breeds and even the smallest of the remaining hatcheries are adding a handful of new (to them) breeds each year. One of the smaller hatcheries here in Missouri actually added Exchequer Leghorns to their breed list recently.

Twenty years ago the rare breed section only ran to just two or three pages in the few hatchery catalogs that were available. At that time the rare breeds were Polish and Brahmas and Hamburgs. Even the offerings of the more commonly seen breeds might number only a dozen or so. White and Barred Rocks, New Hampshire and Rhode Island Reds, Buff Orpingtons, and maybe a handful more were all that were offered except for those listed at a few high-volume hatcheries.

The temptation now is to open a catalog or breeder's list and then order two of these, three of those, and four of something else until the minimum number of twenty-five chicks is met. In short order the bill could reach a couple of hundred dollars and still not be enough of any one breed to launch even a small flock.

The temptation to add breeds is always great, but the focus should be on poultry production and not on being the keeper of a feathery zoo. A few years ago, after a long winter afternoon of chores, I came into the house to tell Phyllis we were up to twenty-three breeds and varieties and something had to go. A year later we were down to seventeen breeds. This number was still too many. Once word had gotten out that we were involved in poultry breed preservation work, birds just seemed to find their way to our doorstep. A pair from here, an old trio from a former 4-H project, and odd hens and roosters enough to supply a soup company came our way. They still do arrive from time-to-time now.

For everyone there should be a very clear purpose for owning any livestock species. Impulse buys should never eat or breathe. With chickens, the production goals are fairly straightforward — eggs, poultry meat or seedstock. Even if your wish is just for a small flock to supply your own table, it should not be a mishmash of birds that leads to mongrelization.

Just about everyone had a grandma or a neighbor with a mixed flock of red, buff, white, and barred birds. They produced some eggs and a bird for Sunday dinner now and again, but there was nothing really sustainable or predictable about such a flock. There is a very great difference between being the owner of a few chickens and the keeper of valuable poultry genetics. Heirloom and rare breeds are in desperate need of those who will keep them bred pure and true to the type standard — very essential for getting optimal performance from them. If your need is primarily for eggs, focus on a breed or breeds bred just for that — good egg production. If your usual market is poultry meat then focus on those breeds with the frame type and growth needed to make them efficient producers of meat. If your focus is to market both eggs and meat, then it is best to keep two separate and distinctly-bred flocks.

The seedstock producer will select choice breeds based on both personal experience and tastes and with the commitment to build on the genetic strengths of those birds. The task is to build on those strengths while retaining good breed character and purity in the resulting offspring. Certainly some breeds will be in greater demand than others and this should be reflected in the numbers kept. For every rare breed bird sold, fifty, a hundred or even more of the traditional breeds will be sold. A great many people still just want a plain red hen with no bells, no whistles, and no complications.

At most poultry marketing events the "old red hen" and the "Dominecker" are still the most sought out and will be forever, I believe. The "Dominecker" is to most a barred Plymouth Rock and not a true Dominique. The red hen may be a red sex-link bird or a "performance" or a "production" red bird and far from any real level of genetic purity.

Sex-linked birds date back nearly seventy-five years and are bred first and foremost for sight sexing at hatching. They are the product of crosses involving the gold and silver color factors or barring gene to create baby pullets and cockerels that will differ from each other in down color or pattern. The most often used pure breeds in these crosses are the New Hampshire Red, Barred Plymouth Rock, Rhode Island White or Delaware. Using a barred male in such a cross, for example, will result in the little cockerels showing a barring pattern.

Such birds also receive a boost in size and hardiness after hatching as a result of what is known as heterosis or hybrid vigor. This increase in size and vigor should go on to serve them well throughout their entire lifetime. Any real boost in factors such as egg production or meat yield will most likely come from strengths already present in their purebred parents. Also, you cannot successfully breed forward from sex-linked birds — it is a true terminal cross.

After a number of years, we started to narrow our focus to just poultry breeds with a strong association with the small Midwest family farms. Some have a bit of a variation from the norm to enable us to give them a bit of our own spin, but this category is at once broad and yet it has great practical roots and economic potential. Perhaps explaining something about our choices may help others with their decision-making processes.

A long-time favorite breed of mine is the Wyandotte. The first heirloom breed I sought to own and to start building a brown egg flock were White Wyandottes. White birds are a real favorite of mine, although many find them too plain or too subject to predator loss. They don't blend well into the background and they really stand out on an open roost on a moonlit night.

Now we keep Blue Wyandottes, as this is a most popular color. The feathering and rosecombs make them quite winter hardy in our simpler housing. Color is a consideration and should reflect buyer preferences. Our main flock, although still quite small by any standard, is a set of Rosecomb

Rhode Island Reds. They are the true old-school variety, dark mahogany color. The importance of Rosecombs was driven home for me several years ago when a hard March freeze caught us with all of our breeding groups out in simple pens around the poultry yard. Comb damage was widespread on the single combed varieties and with some of the breeds we didn't get fertile eggs until well into the month of June.

Freeze damage to the comb and wattles and the resultant period of infection and low-grade fever can render the birds sterile for quite some time, especially males with their larger combs and wattles. On a cold night these males will not tuck their head under a wing like a hen will. That year our Rosecomb varieties suffered no such setbacks. The Reds are also the breed of my youth and red is the color most favored by my wife, Phyllis.

Buff birds always seem to sell well and for most people "the buff bird" is the Buff Orpington. This is despite their lack of yellow feet and skin that is so popular with American consumers. We have worked with both the Buff Wyandotte and Buff Rock breeds, but when offering chicks for sale we noticed a fairly high degree of consumer reluctance when they were told that our buff chicks weren't Orpingtons.

Acknowledging that the consumer is always right, as they have the money — we have begun forming a small flock of Buff Orpingtons. We are going to emphasize a bit of size and increased production in this breeding group. We will be looking for select birds to add to this program for some time to come. Even with breed preservation work, economic factors must be considered in breed selection.

In the heritage movement I have felt that the white egg-laying breeds have been neglected for far too long. The Mediterranean class of birds is especially good at laying large numbers of eggs and does so with relatively modest space and feed needs. As for challenges, they are often high metabolism birds and some lines can be rather flighty. Still, among our valued family pictures are photos of my grandparents with their White "English" Leghorns from the thirties and forties.

These birds were not the White Leghorns of today that are cage bred, but were large birds of a most substantial frame. They helped my family weather the waning days of the Great Depression and the often protein-short years of World War II. I would love to run into their like today and would gladly make room for them in our poultry yard. Our choice of their kindred is the Exchequer Leghorn, the Scottish Leghorn. The breed should be mottled, although predominantly white in color. They are one of the largest of the Leghorn varieties and have a very calm nature. The breed needs work on size and color and there is a problem with maintaining a clear yellow leg color, but their serene nature has us hooked.

Other than the Exchequer Leghorns perhaps the "rarest of the rare" on our farm are the Rhode Island Whites and we have them in both Rosecomb and Single Comb types. The Rosecombs are the only one of the two now recognized by the American Poultry Association and they are available only from a handful of breeders. The first hatchery flock to be launched in modern times has only just been started during the summer of 2006. The birds are a bit slow to grow and they do take time to develop. They are fair layers, but where they really shine is at about sixteen weeks of age when they really take on fill and become very solid and well formed.

The Single Comb Rhode Island White came into our lives several years ago in a shipment of chicks from a heritage breeder to the north of us. They are a bit smaller then the White Rock, but I have yet to find a brown egg layer of their equal. They lay earlier and longer than anything else on the farm and after the first four or five eggs do not lay pullet eggs. They are said to have a completely different genetic background than the Rosecomb Rhode Island White variety.

I have to confess that I have dropped this variety twice for something flashier in appearance, but the Single Comb Rhode Island White has become quite common in our area. I come back to them because of their productivity and have to believe that this is a variety that just needs a chance and will make a lot of friends for itself.

Like every small farmer we always have an experiment or two going on among the pens and coops. In our "lab pen" now is a red barred bird that was plucked from a set of hatchery run Barred Rocks. He is being mated with White Rock females to clean up and brighten the red barring in his resulting offspring. It may or may not work just as we hope, but it is worth the try. Red and Buff Barred Rocks were once both bred in the United States and these are color patterns that should prove quite popular.

I used the term "hatchery bird" above and one of the factors to be worked through in today's burgeoning poultry scene is what might be called the "hatchery conundrum."

I can remember a time when nearly every county in the Midwest had a hatchery, sometimes two. Throughout the nation were a great many very large hatcheries that might offer literally dozens of breeds and varieties. Over time both groups began to decline in number and now a relative

BREEDING HOMEGROWNS

Shady Lane Poultry Farm's owner, Matt John is a big fan of American heirloom poultry breeds and raises about 40 varieties, many of them with a long homegrown history in this country. Included among his diverse flocks of poultry are nine that he calls "heirloom and rare" varieties. Matt likes the American heirloom breeds because they have a homegrown history and he feels they are best suited for this country. He breeds his poultry for trueness to the standard type and productivity. He plans to soon expand his breeding criteria and will evaluate his heirloom and rare breeds for their shell quality characteristics (something that is done less so at present for the rarer breeds than with the industry breeds).

For 2009, Shady Lane Poultry Farm's most popular heirloom and rare breeds are Black Java, Buckeye, Delaware, and Dominique. Still Matt comments that only small numbers are being grown of most of these breeds. Matt has noticed that there is an up tick in rare breed interest after a specific breed is mentioned in an article. Like the fashion industry, heirloom poultry breeds also change in popularity with each passing year. Whatever the type, the heirloom and rare breeds that Matt John raises are dual-purpose birds that have a good compatibility with their environment.

Shady Lane Poultry Farm has been in business for seven years selling day-old chicks. Shady Lane Poultry sells the chicks from January through March. The day-old chicks are shipped throughout the continental U.S. through the Post Office, and Matt has had no trouble shipping chicks this way. While Matt does collect a few eggs for his own family use, they do not offer them for sale commercially.

After this shipping season closes, the next year's replacement breeding stock are brought together and allowed to pasture. Matt believes that the healthiest and best way to raise and condition chickens is by letting them range outdoors. The breeding stock molt in the fall and are then separated into breeds. A rooster is introduced to each variety so they begin producing fertile eggs for hatching. During the colder months the adult breeding birds are kept inside under lights to better control the lay period.

In 2008 the business was moved from Winchester, Kentucky to their present home in Columbus, Indiana. One thousand birds were moved to Indiana with Matt John, a 180-mile journey that would take three hours non-stop by car.

The move had planned and unplanned consequences. It was planned that no chicks would be sold during the year after moving as the flock was being rebuilt. But an unexpected setback from a major flood did cause additional delays in getting the business reopened.

Matt John is the owner of Shady Lane Poultry Farm in Columbus, Indiana. He can be reached at info@shadylanepoultry.com.

handful hold sway over the poultry community. Today the large poultry corporations produce their own replacements in house.

Some of these hatcheries still offer a great many breeds and varieties, but a relative few own anything in the way of breeding flocks. Most hatcheries buy at least a portion of their hatching eggs from independent suppliers and, for some breeds, just one producer may supply many hatcheries.

Several years ago I ordered a set of White Wyandotte chicks from a Southwest Missouri hatchery. Later that year, wanting to add some birds of different breeding, I placed an order with another hatchery several states away. It was a hatchery still shipping chicks after most Missouri hatcheries had stopped shipping for the season. Imagine my deflated feelings when my "new blood" arrived in a chick box with a postmark from the same town where my first order was sent.

Many have also reported great variability in quality when ordering from some hatcheries. Essentially, hatching egg suppliers and hatcheries are said to fall into one of three groups depending on the amount of attention they pay to type quality. Some of the country's top breeders sell hatching eggs to nearby hatcheries, then there is a middle group that produce birds that can be said to be average for the breed, and a few seem happy if the colors just match.

Do not just blithely mail off a check to a hatchery assuming you are going to be sent show winners and true stud mating stock.

SELECTING A HATCHERY

There are many factors to consider before placing an order.

Read the catalog or breed list carefully and then read it again and again. Carefully note all descriptions and whether birds are offered in different grades. Beware of birds denoted "not for exhibit."

The rare and heritage breeds are generally going to be available in truly limited numbers regardless of the supplier. Prices that range up to seven dollars per chick are not unusual.

Take the time to ask detailed questions about egg origins and the breeding behind the supplying flocks.

Smaller hatcheries may offer fewer varieties, but don't disregard this source as some of these have been working with their own, long-established breeding lines.

Hatchery flocks are generally turned at the end of each hatching season to trim costs and assure good levels of egg production during the primary hatching season from February through June. Some people have reported acquiring good adults at the time of these seasonal dispersals.

In the off-season, mid-June through early-January, many chicks are shipped from one of just three large, southern-based hatcheries regardless of where the chick orders are placed.

If exceptional chicks are received, make a note of which week of the month they were shipped. Then future orders can be placed for that same week of the month with the assumption that they too will be from the same source flock.

Consider ordering from smaller hatcheries that are somewhat off the beaten path. They are more likely to be working with their own breeding lines, lines that may be long established, and that are of the classic breed type.

Directories of hatcheries both within your home state and nationally should be available from your local extension office or state department of agriculture. Request directories of those producers participating in your state's Poultry Improvement Plan and the National Poultry Improvement Plan. These are listings of the breeders and hatcheries participating in the testing program for the diseases Pullorum and Paratyphoid.

I would not discount an independent producer who did not participate in this program, however. The hatchery conundrum is magnified further by the role that those big, multicolored hatchery catalogs have come to play. They certainly aren't poultry bibles, but near countless times I have seen greenhorns roaming bird swaps and farmers' markets with one rolled tightly in hand. They treat a bird's picture in the catalog as some sort of acid test of its bona fides. In these catalogs you will not find the rarest of breeds. At best they offer a thumbnail sketch of the breeds they do offer and many of their illustrations can be quite misleading. Many of those breed illustrations are taken from various sources and depict artists' renditions of birds edited to display an idealized, perfect type. Let me cite just two anecdotes about catalog-schooled chicken shoppers here. One Spring I was set up at the local farmers' market with a good set of Blue Wyandotte chicks. A young 4-H'er stopped by several times with a certain hatchery catalog in hand. I took the time to tell him about our Blue Wyandotte breeding program. During his last stop he told me, "Mama said they can't be that. They're not in the catalog."

The year before, at the same market, a very angry woman set down a quite good coop of White Crested Black Polish and Blue Splashed Polish, a great many of them were pullets. She was mad, by thunder, because she had ordered White Crested Blue Polish and just look what they sent her! She wasn't willing to listen to any explanations although several well-intentioned poultry men and women there that day had tried.

Quickly then, the birds were snapped up by veteran producers who understood the ins and outs of blue poultry breeding. They bought them for a price not much more than what she had given for them as baby chicks. Those black and splashed birds when mated together would produce all blue chicks. The lady hadn't even bothered to read the captions under the pictures in her catalog.

At this time the American Poultry Associations' *Standard of Perfection* and many sales brochures and catalogs are illustrated with an artist's rendition of the ideal type bird. Such birds will never be bred and such paintings and drawings lack the detail and color intensity of photographs produced by professional livestock photographers.

It is not always clear how idealized these depictions are. Some catalog pictures are taken from real life and do show the variability that can and does occur in any hatching. Even birds that would fare poorly in showroom competitions can have a role to play in a well thought out breeding program. With some of the truly rare breeds, virtually every bird hatched may have a breeding role to play until viable numbers are reestablished.

I am certainly no artist, but I do believe that actual photos of birds with real merit would better serve all those from rank beginners to veteran family farmers. These photos can also show their flaws and make them the kind of teaching aids and promotional materials that are honestly needed. At present those putting together meat and laying bird flocks find the photo illustrations from earlier copies of the APA *Standard of Perfection* to be more useful than later versions. Breeders concerned with having both winning breed character and good, productive type drew up those earlier type standards.

Interestingly, in the past few years, printed poultry materials including those hatchery catalogs from yesteryear have become pricey collectibles. When a new chick catalog arrives, things here come to a complete stop until I give it at least one quick run through. Yes, I will still order from a hatchery from time-to-time and I believe there is nothing wrong with trialing a new strain or breed every now and again.

I know of some veteran breeders who make it a policy to order out a few chicks of their breed each year from a different source and some of a new breed (for them) just to see what is now available. This usually turns into a rare learning experience with a bit of fun attached to it.

The heirloom birds are the glamor girls of the moment, but the reasons for making a start with them are much more than mere exotic looks and hype. The heirloom birds have held on for lo these many years in large part because they are right tough birds. A little Hamburg hen I saw on public display sometime back was evidence of this. She was of one of the rarer color varieties and when I saw her she was starting to show her years. When I commented on her not looking up to a trip to town her owner proudly defended her.

When he had located her she was one of the very last of her kind in the whole world. Even at her advanced age she had still bred on and contributed to the comeback of her kind. With that in mind I had to admit that even with several feathers gone she was still quite beautiful.

It is this grit and vigor, an ability to work well all around the country, and in all types of facilities that really makes them valuable. As range producers are discovering, you don't just pull cage layers and hothouse broilers out of confinement units and make them work anew in the great outdoors. It is not mere serendipity that they are coming out of a decades-long stasis at a time when a more natural course of poultry production is again unfolding. It could be said that these birds are again ready to meet the destiny for which they were created. It is time for them to be taken up by a new generation of artisanal producers. Such producers have more in common with the fine vintner's mentality than the corporate thinking of those like Tyson or Perdue.

However, it is a mindset that must be tempered with a bit of business-like thinking. A half-dozen hens will generally produce enough eggs for the table needs of a family of four. Thirty heavy hens at eighty percent production and with eighty percent hatchability will produce a lot of broiler chicks over the course of a hatching season. One reflects the mindset of a subsistence producer and the other a smallholder working with a business plan.

One hundred layers at about eighty percent production will produce fifty dozen eggs a week. At $1.25 or more a dozen they will produce a nice little weekly sum. With a bit of extra management and marketing they can bring in two or even three dollars a dozen for natural, free range, additive-free eggs. They are on the way to being truly fancy table eggs.

Take them into a city of some size and they may become four dollars a dozen organic fare. Even at those prices, pound for pound, they are still quite inexpensive sources of high quality protein.

They have their greatest value only if a plan of production is matched with a plan for marketing that is designed with a focus on high value. At our Missouri farmers' markets today the darkest brown eggs are selling regularly for $2.50 to $3.50 per dozen. As a result, in our local area, there are several Welsummer and Maran laying flocks building as quickly as possible to the one hundred fifty to two hundred hen level. Even with these numbers reached there will still be a need for a lot of selective breeding for improved laying performance. With only a modest start, such a flock can be started with the purchase of just a single trio of breeders and the two hundred number reached in the fourth or fifth year.

A broiler breeding flock of White Rocks, Delawares or White Wyandottes can be launched with a single box of twenty-five as hatched chicks. It will take time to achieve, but if they are well bred and of good type they will produce twelve-week broilers in due time. It will take a lot of pretty ruthless culling and it may even be necessary to completely start over a time or two, but the reward is a self-perpetuating poultry venture for all ages of the farm family.

With a production goal firmly in mind, the next matter to be resolved is that of breed selection. "What breeds are best for the family farm?" is the question I am most often asked. Fortunately, for what little desire I do have to be politically correct, it is a question that has no correct answer. I can offend no one!

To those really wanting to keep things simple there are a handful of selection criteria to consider when selecting a breed or breeds. These are rosecombs, clean legs, solid colors, yellow flesh and yellow legs. The rosecombs are an aid in cold climates, clean legs are easier to manage on pasture and dirt lots than feathered leg birds, yellow skinned legs are preferred by the majority of United States consumers, and the solid colors are far simpler to manage in a breeding program. Black and white are by far the easiest colors to breed although there are actually five different white color variants in chickens ranging from a silver white to the absence of all colors.

However, the simple truth is that with any livestock venture you will encounter the greatest success only when you are totally invested in it — heart and mind. That means that you must be working with the birds that appeal the greatest to you and that will challenge you to do an ever better job with them. If your heart is set on Russian Orloffs, Light Sussex or Golden Spangled Hamburgs, then go for it, but be willing to accept the limits that naturally come with such breeds.

The cage-free egg is the poultry pearl commanding great price at the moment and to many folks, the browner its shell, the better. However, the greatest number of eggs with the lowest costs to produce them will come from some of the white egg layers. Although few in numbers, heritage breeds with good egg production histories would have to include the Anconas in both comb types, Minorcas in both combs, and some of the colored Leghorn varieties (also in both combs). Logistics also favors starting with just a single breed and keying upon it as you develop skills and experience. There is much merit in this approach as that one breed then receives the fullest focus and breeders become a real student of it. Gradually they can build numbers and breeders can increase facilities as they gain that experience.

Still some of these rare breeds exist in such small numbers that starts are made with a scant handful of birds and it may be years before any sort of numbers are achieved. I waited nearly two years for my first trio of Welsummers to arrive. We got them young, late in the fall and then lost one of the pullets early in their first breeding season. An important first lesson to learn is that good breeders don't necessarily need large numbers to be successful.

This point is illustrated with a rather classic anecdote. It is said that good breeders tend to fall into one of two groups and those groups are defined thusly. If asked to ready thirty good young birds for a buyer, the members of one group would hatch and rear thirty youngsters. The other group would hatch one hundred or more birds to be assured of getting the desired quality in the desired number. I will not say if either group is better than the other, but the former should be recognized for taking years to develop the quality and expertise to operate at that exacting of a level. Consistent, precision performance generally comes only after long years of dedicated, good effort.

An operation based on one pure breed is probably the definitive standard for the calling, but there have been other approaches to poultry keeping — many of them with long traditions of their own. Many have kept multiple varieties of a single breed. Others have kept separate flocks of meat and egg breeds. Many keep several breeds of a single class, such as American or Continental. Poultry keeping also has a long tradition of what were called "string men." Part trader, part showman, and part breeder, they kept numbers of several breeds to exhibit and offer for sale. They had a valuable role to play in the preservation and advancement of any number of breeds.

That question of numbers is just one more to be resolved by the heritage advocate. The best answer to it is to operate only at the level at which you feel most comfortable and produce only those numbers to serve a market that is operating at premium prices. The cheaper you are willing to sell a product the more of it you will sell, but at the expense of destroying a premium market and both your current and future profit margins. It makes no economic sense at all to try to match a two-dollar per bird profit margin with a 25-cent per bird margin on eight birds just to be a big producer.

If you can get five hundred broilers sold each year at $1.80 per pound for four pound dressed birds, why pare down prices by twenty cents a pound just to get a few more sold? The small-scale producer should take his or her rewards from optimal production rather than high volume output. The markets for top quality poultry products can absorb only so much due to their generally local nature and their dependence upon direct marketing and producer services. Consumers that want just any chicken can always get one from the neighborhood supermarket. Top quality poultry producers can provide so much more with their eggs, meat and extras.

A poultry venture should be only of sufficient size to fit neatly into the farm's venture mix and not come to dominate it or skew it too much toward one category of production. That mix can be broadened with multiple, different themed poultry ventures that complement each other, but the farm should never become all feathers all of the time.

Time and experience has shown that it is all but impossible now to create a single farming venture with the yearly net earning power of twenty thousand dollars or so (a good, full-time income for one person in many areas). Ten small ventures netting just two thousand dollars each will generate that level of earnings. They are much easier and less costly to create and should one or two fail entirely and need to be replaced — the whole is not lost. In fact, they are much easier to fine tune, revamp or even remove and replace completely.

Seek to grow the farming ventures by no more than five hundred to one thousand dollar earning increments and they can perhaps be expanded upon for a number of years before reaching the saturation point for locally available premium paying markets. Locking in and holding a profitable level of production is a most valid business practice. Remember, growth simply for growth's sake is the philosophy of the cancer cell.

As noted earlier, those working with the heritage breeds may be best served by keeping separate breed flocks for separate purposes. One breed flock for eggs and another, probably much smaller flock, to produce broilers and roasters would be a norm. The seedstock producer may have a number of small, select flocks of breeds noted for different traits. Some might even try to fit another poultry species or two (waterfowl, guineas, turkeys, etc.) into their enterprise mix.

There are individual breeders advertising scores of breeds, but often they are kept in small breeding groups of no more than a trio or a breeding pen of just four or five birds with a rotation of roosters. To be honest, too great a number of breeds can raise some troubling questions to potential buyers and can spread thin even the most hardworking of producers. In a large operation, a single rooster loose for even an hour or two can raise hob in a number of breeding pens.

Realistically, it is going to be much easier now to put together a flock of several hundred good Rocks of the more common varieties then even a trio or a pen of five good Mahogany Faverolles or Crele Penedesencas. Such a flock may better serve the needs of some small-scale producers, too.

The heritage breeds have been here and have been waiting for, quite literally, decades on end. They fell out of favor through no fault of their own. Each breed had a role to play then and can fulfill a similar one now.

Their rediscovery, while not a reinventing of the wheel, does amount to getting that wheel started rolling again. Thus the different breeds will be starting from different points and from different roles. The performance of all can be ramped upward, but each will have a different performance curve shared by what was bred into it originally. The Hamburg, once called "the everyday layer," will have potential for increased number of eggs, but egg size will be limited by the bird's modest size and its original role as an efficient producer in a modest environment.

The breeds in the heirloom sector have had a long history as simple farmers' fowl. They were expected to do their part to scratch up a living both for themselves and the farm families that owned them. In contrast, the bird bred for confinement is also bred to have a lack of vigor and simple hustle, whilst they live inside a temperature-regulated building where feedstuffs are just dumped before them, there is no need for durability and toughness. These qualities may even be counterproductive in the confinement environment.

Yet heirloom breeds as vividly colored and spectacularly feathered as the Light Sussex and Light Brahma were once bred to very high performance standards. The interest in heritage breeds began around breeds such as the Dominique that were deemed truly historically significant. Successful preservation work will one day again place such birds in a practical light. Preservation will build numbers, distribute the birds broadly, see them returned to type standards conducive to improved economic performance, and bring them to greater attention both in the countryside and the marketplace.

Another important factor weighing in favor of keeping heirloom breeds is the simple fact that they are purebred birds. They epitomize sustainable agriculture in that once a purebred flock is established it then becomes virtually self-sustaining. Unlike the hybridized laying and broiler varieties of corporate producers, heirloom genetics are still in the control of, and reproducible by, independent producers at all production levels. The loss of control over the seedstock is the one occurrence from which independent producers can never recover.

Once in position, a purebred flock can be shaped and selectively bred to provide optimum performance on individual farms and in specific service to various niche markets. Through selective breeding for strength in these areas, egg production, days to market, and muscle yield can all be substantially improved. There is measurable evidence that these so-called "performance traits" actually respond better to selective breeding than any other group of genetic traits.

A Black Giant cannot be bred to lay like a Leghorn or a Leghorn bred to produce a broad breast. Such would not be in keeping with their breed character or with the purposes for which these breeds were developed. With broad-based populations there is always within every small, purebred flock, the seeds for any needed change and an assured future for both breed and breeder. Individual breeders will always vary enough on matters of type and form to assure that the breeds will retain a fair element of genetic vigor — the vehicle for change as needed.

The industrial model has brought to poultry production a cookie cutter mindset that sees little or no value in genetic diversity and richness. "Pack 'em into the houses, plunk 'em, pluck 'em, and parcel 'em out by strips and nuggets," has been their sole mantra.

I don't believe that there is something that could or should be called the heritage or heirloom poultry business. In point of fact, if our preservation efforts do bear ample fruit, such terms should become obsolete.

What is before us now is a season of opportunity. For the moment, breed preservation is the foremost task, but coming out of it should be an ongoing role for this and future generations of independent poultry producers. Purebred stock offers them the means to control their own destiny by producing their own seedstock and shaping it to meet the needs of their farms and markets.

The new and rare aspects of many birds are very much a part of the heirloom scene right now, but we cannot build upon the practice of fad chasing. Ten years ago few knew what a Welsummer was, five years ago a good trio could set you back three figures and even now, day-old Welsummer chicks are five to six dollars each in some catalogs. Yet at our local farmers' market it is not uncommon to see twenty to thirty Welsummers cooped up and offered for sale on some Saturday mornings.

The first of the heirloom stars was the Dominique, as noted earlier. Now they are a staple in nearly every hatchery catalog. Dominiques have a new breeders' group and are seen at many poultry shows. They have been followed in fairly quick order by a number of other "hot" breeds such as the Dorking, Sumatra, Welsummer, Maran, and now, the Penedesenca (the dark brown egg layer from Spain).

Due to their prolific nature, each hot breeds moment in the sun is rather short-lived. It can be an exciting time with some per bird prices not having been seen since the glory days of the old Madison Square Garden poultry shows. When this fire burns down, hopefully each breed will be left with a group of core breeders committed to a long-term future based on a concern for quality and the good of the breed.

This is a truly remarkable time for those of us who have had an affinity for chickens. A time many of us thought would never come again. However it is a time not without obstacles and risks.

Some of the breeds have been left sadly neglected and some are holding on with numbers so small that they might be lost with a single misstep or moment of thoughtlessness. Rare breeds aren't for everyone and some of them were of only minor importance even in their heyday. To endeavor to build a flock or flocks is to take on the work of a lifetime, but then it was ever so, even during the era of the stockman's calling.

I have taken a rather stern tone here and hadn't really meant to do so. It is crucial, however, to emphasize that some of these birds are still in rather perilous straits. Producers falter, birds are dispersed never to be retrieved again, and the role of small farmers is diminished just a tad bit more. As a result of issues like Avian Influenza, the national animal identification systems and other challenges now facing poultry keeping, the current image of rare bird hobbyist or mere dabbler serves neither the birds nor the small farm community well. Those serious about heritage breed propagation need to be building flocks with a true mission statement and a lifetime of quality improving goals. Next year's chick crop will be better than the one at this year's end. Each flock should be a year stronger and a year deeper in the traditions and genetic values on which the breed was founded.

HERITAGE BREED CHICKENS

FINDING THE BEST FIT FOR YOUR FARM

There are numerous breeds of chickens available to the modern day homestead. Of course what you ultimately choose will depend on many different factors! It seems a simple thing, to choose a chicken breed to raise but there is so much difference in breeds, quality, purpose, and location that it is not as cut and dried as it would seem. Within the various breeds are areas like egg production, egg color, temperament, meat production, broodiness, and survival skills in various situations, as well as personal preference for a certain coloring or pattern. So, then, the first thing is to identify why you are buying the poultry in the first place. Will you use them primarily as bug control, egg production, meat, or show? Do you want chickens that are more aggressive, say, if a coyote comes into the yard? Or, would you prefer a breed that is more gentle around children and needs more protection?

Does the look of the chicken make a difference to you? Do you have a lot of hawks? Traditional white chickens will get picked off much quicker by a hawk than will a dark patterned bird. Docile hens, like Buff

Orpington, will cower down when a predator stalks them rather than try to seek shelter. A more aggressive breed, like a Dutch, might fare better with a dog but also chase your children pecking at their legs!

HERITAGE BREEDS ARE DUAL-PURPOSE BREEDS

Birds that are good egg layers are often not good meat producers! However, heritage breeds are not genetically engineered to be specialized and so lay eggs as well are produce tender, delicious meat.

Leghorns are kept for egg production. They lay white eggs. They are able to forage for themselves and so are good for free range situations, although they do not go broody as well as some of the other breeds. Basically this means that they are not good about hatching their eggs if you want to raise chicks on your homestead. They also don't produce a lot of meat.

Bantams lay tiny eggs that are the delight of my smaller children. Two of these minuscule eggs fried and on a plate with a toast triangle is a magical breakfast. The eggs are often colorful pastels. These birds are small and make good pets or show birds for children.

Plymouth Rock is a heritage breed. As with most heritage breeds you will find that it is multipurpose. It is a good egg layer, a good brood hen, and produces a fair amount of meat. It is a docile breed.

Rhode Island Red is another heritage breed that is dual purpose and lays abundant numbers of eggs and is probably one of the best dual purpose breeds for a small homestead.

Delawares are excellent egg layers, a good dual-purpose breed. They are listed as critical on the American Livestock Conservancy list.

Holland is another breed on the critical list. This is currently one of the rarest of the heritage breeds and one of the few that lays white eggs.

Wyandottes are not particularly rare, however they are a heritage breed. I add these because the last chickens we raised (we now have Barred Rocks) were Golden Laced Wyandottes and they were beautiful!

There is, of course, nothing in the world like the taste of your own farm fresh, organic eggs and meat. Chickens are an easy way to begin food production on the homestead and work a little closer to self-sufficiency. Be sure, when you are ordering your chicks, to have the vaccinated and then use nonmedicated feed. In this way you will not have to worry about residual medications in the eggs or meat. It is very relaxing to sit and watch chickens and for me, they are what puts the *home* in homestead!

Marye Audet is a freelance writer, blogger and photographer specializing in food, green issues, and homesteading. She is currently living in Texas and restoring a 100-year-old farmhouse.

·CHAPTER 3·

MAKING YOUR START

SINGLE COMB WHITE
• LEGHORN MALE •

SINGLE COMB WHITE
• LEGHORN FEMALE •

DOMINIQUE
• MALE •

DOMINIQUE
• FEMALE •

OUR FIRST HERITAGE BREED PURCHASE WAS A BOX of twenty-five White Wyandotte chicks and our first rare breed acquisition was a trio of Welsummers. At the time each was our only means to access a start of those particular breeds.

The three primary means of starting with chickens are hatching eggs, chicks, and started or adult birds. Each option has its pluses and minuses. In most instances a poultry flock is launched using at least two and generally all three of the above.

Long before the acquisition of even a single bird, flock creation begins with a number of preliminary steps. Each is designed to prepare the producer and his or her facilities for the selection and rearing processes. The first task is, simply, assembling as much information as possible about the breed or breeds of choice and then researching available sources. The American Poultry Association and various breed groups have done quite a good job of documenting the breeds and varieties they sanction or represent. Breeds not so sanctioned in this country may be so in their countries of origin. Descriptive data on such birds may be obtainable on the Internet or by contacting government agricultural authorities in those countries. For assistance with the latter, try first contacting their embassies in the United States.

For guides to what should be beneath the feathers — those type factors conducive to egg and meat production, turn first to the teaching texts. The pre-war M.A. Jull texts are still good beginners' books and something even veteran producers will find useful, too. Often state universities publish guides and pamphlets on poultry selection, feed suppliers sometimes offer some good beginner pamphlets, and there are now a great number of poultry-themed books (both current and vintage volumes) available. You should strive to be ever current in your fields of endeavor and this means reading extensively from available magazines and newsletters. A great many general farming periodicals are again broadening their poultry coverage. Your local Extension office can get you pointed in the right direction for a great many current publications in the field. The Internet is alive with chicken information and poultry-themed chat rooms. Much of it is rock solid information and in those discussion groups you will encounter many deep discourses by those who have earned the right to their opinions. Unfortunately, the Internet is also alive with much false information and self-promoters. In some of the chat groups you will encounter some real debates on both issues of the day and ultra fine points of breeding and management. Participate and enjoy, but just don't take everything you find there as gospel. Also, beware of what may be little more than thinly concealed marketing ploys. Do your homework carefully and ask lots and lots of questions.

A part of becoming well grounded in your breed or breeds of interest is to form mental images of birds of the most desirable type and those that you need to add to upgrade your own flock. These images can be formed after careful study of the American Poultry Association breed standards, birds in exhibits, and birds in the home flock. With such images in place the selection process then becomes a matter of comparing birds at hand with those carefully formulated mental images.

There is really nothing mystic about the so-called stockman's eye. It is a matter of training and taking the time to really look at the bird or birds before you. Learn to see them as a whole, but a whole assembled from a number of key and analyzable parts.

This is a selection measure employed with nearly every livestock species and it is one of those unspoken things that create an image of professionalism. As most veteran producers will tell you, the very best bird or animal in a pen, flock, showroom or ring can be detected quickly. The real time is taken up with evaluating those of lesser quality. Where placings in a show go deep, watch the struggles even a veteran judge has with determining the difference between the eighteenth and nineteenth place birds in a class. This leads to a second, very important point in the selection process. With the rare and heirloom varieties there are often quite limited populations from which to draw. And often such birds may be suffering from some real quality issues. Accept this one hard fact, that with many rare and heritage breeds you are simply going to have to start and work with what you can get. Certainly, you can't consider birds with severe physical defects, breed type disqualifications or health problems, but just about anything else goes. Once you get viable numbers built you can begin phasing out birds with problems of type and color. Nor, even with the most widely available of breeds, can you expect breeders to sell to you from the heart of their flocks, not unless you are willing for a lot, an awful lot, of money to change hands. Plan on working with your fair share of older hens, one-eyed roosters, and a great many birds with "color issues." You will be, with many breeds, assembling bits and pieces of genetic material for a great many years. You may have to work with birds that are quite closely related, advanced in their years or that have been otherwise adversely impacted. With some you may get only a handful of chicks raised in the first two or three years.

With very small, endangered populations each and every bird will be crucially important and have a role to play in the always-critical numbers-building stage. You may well find yourself using a bird or birds that in just a few years down the road you would cull out of the flock at a very young age. In a few instances you may even have to employ a complete outcross to keep a small population viable and reproducing, but more on that later.

As a dedicated producer aspiring to a higher level of professionalism you will be about creating your own breeding line or lines. These will be largely linebred and may even trace largely back to just one individual of exceptional merit. Quite often that individual will be a hen although we now have a Rosecomb Rhode Island Red rooster that will grow old with me.

Before the beginner is the option of plugging into an existing breeder's program or combining genetics from two or even three different established lines to create a new breeding line. Both options will require you to take the time to learn what is available, how such lines are performing, and how they might perform in facilities comparable to your own. Good seedstock doesn't come cheap. It may be the work of a lifetime for some breeders. Still, nearly all producers worth their salt will welcome your sincere and sensible questions. Place your phone calls in the early evening hours (remember time zone differences), ask if you are calling at a bad time, and always send stamped, self-addressed envelopes when making mail inquiries. Budget plenty of time for this shopping process and begin with an orderly plan of exactly what you want, a budget, and your list of questions in hand. A budget is most important and it is always better to sacrifice acquiring in numbers and spend more for quality. In both the long and short runs two of the good kind will always do you more good than a hundred of the other.

Birds from a private source, while not being the producer's very best, should be related to his or her best and deemed to be of some real potential. They should be out of a well-proven genetic background. A bird not of award-winning purple ribbon caliber itself can still have the genetic wherewithal to produce birds better than itself, and often far better when paired with other quality birds.

One of the potentially better buys may be some of the older birds being phased out of a flock by their own offspring. Another of the positives of buying from individual breeders is the opportunity to buy chicks or hatching eggs that are akin to some of their very best, but that are not too closely related themselves. When ordering from a commercial hatchery it is nearly always best to assume that everything in the box is a full or half-sibling.

Full sibling and half-sibling matings are the closest possible matings to make and are seldom advised. To prove the strength within a line, however, the mating of a male to his full and half-siblings would be the definitive test. An established breeder should be able to offer birds that are less closely related, such as cousins, in a line. Ask that eggs or chicks of different breeding in a shipment be so marked.

As a youngster still in high school I had the opportunity to begin a correspondence with noted Canadian Rhode Island Red breeder, Mr. Maurice Wallace. Mr. Wallace imparted much wisdom and advanced some thoughts on breeding from an established line that I believe most sound and would like to see tested further.

He proposed that one breeder partner work with another some distance from him or her, in a different climate or environment, and even working with a different type of production facilities. The first breeder would then send the second a respective group from his flock to form a new breeding population of that line but one that will be facing the challenges of life in a new and different environment. The second producer should then breed from that new population for a number of generations.

After a time of facing a new environment, different management style, and even different feedstuffs, this line should have evolved different from the original as a result of the different challenges faced by the birds. They will still be quite close genetically and should look the same in appearance, but they still will be somewhat dissimilar due to the different challenges to which they have had to respond. Birds produced by the second producer would then be returned to the first to be added to the ongoing linebreeding program there. Good breed type would be reinforced from both sides of such a mating along with an added shot of vigor going to the offspring due to something akin to heterosis or hybrid vigor. The challenges the birds had faced and bred through should make them just genetically dissimilar enough for such a thing to occur. Thus, two cooperating producers could do much to keep a linebred group of birds both prepotent and more vigorous.

Some of the heritage breeds have been kept alive by only a bare handful of breeders, who, even if widely scattered, may be working with birds that are from a single, originating source. Your only option then would be to plug into an already rather closely aligned breeding population. The second option of creating a new line by drawing from two or three existing lines will certainly take time and careful execution. What is that old punch line? You can have it fast or good, but not both. It is not exactly a "two steps forward, one step back" thing, but it could be the endeavor of a great many years. It will take considerable time to breed in the desired levels of quality and predictability into the line and the matings produced from it. The first challenge will often be to just find good birds from two or three genetically distinct and strong sources. Two may be better and you certainly don't want to try to mix too many ingredients into this genetic stew that you are creating for your poultry flock.

Evaluate the resulting breeding birds and their offspring most carefully with an eye to breeding strengths to strengths and always selecting from the most vigorous and well grown of the birds. The matings should be small, closely monitored, and culled ruthlessly. While recombining some of these birds it is best to continue breeding some of them true to their existing lines. They serve both as a control and a pure resource, which you can go back to in order to create more combinations. Your task is to build by what old-timers term "breeding strength to strength." You take the birds from each group strongest in how they manifest desirable traits and then breed them together to further amplify those traits. Thus you should create a line prepotent for these good traits and that will breed on consistently. While doing this you must also seek to maintain consistent levels of libido, vigor, and will to grow and flourish.

The producer with a purebred flock producing basically table eggs or slaughter stock, sometimes termed a "commercial" producer, can practice a stud-type of mating. After a careful analysis of the strengths and weaknesses of the existing flock and its offspring, new males strong in the areas where greatest weakness are noted may be bought from outside sources to be bred into the existing flock. In a seedstock producing flock the need may sometimes arise to bring in an outside bird strong in certain traits, too, but that will be discussed later.

These males should then be bred to a few hens and pullets on a trial basis to see if desired results are obtained. If these mating trials prove out ("nick"), then the male and one or two of his best offspring can be used across a greater segment of the flock. Always retain one or two matings pure to the line if at all possible.

For best results focus on upgrading only one but never more than two traits at a time. Trying to do too much at one time in the breeding pen can result in losing ground in one area while boosting performance in another. To upgrade performance in some areas this type of selective breeding may have to be continued for several generations. The impact of a good male will be felt primarily through his daughters and granddaughters that go back into the flock. In the commercial flock the male is always fifty percent of that flock and its short-term future performance.

As noted earlier the three options for acquiring new or replacement seedstock are hatching eggs, baby chicks, started or adult birds. There is no all good or all bad with any of these actions. They are probably best used in combination, and with the very best birds even cost factors are quite similar.

There is a long history to the hatching egg trade and among my grandparents' antiques and collectibles were a number of hatching egg shipping crates. One was made of one half inch by one-half inch hardwood slats fitted together to form an open-weave box that would accommodate several flats of eggs. It had a solid wood bottom and fitted top and a wrought iron handle for carrying. Another had a box made of aircraft aluminum and had egg dividers made of rigid cardboard with metal reinforced top and bottom edges.

Hatching eggs were once shippable when live birds weren't and that happened again one autumn just a few years back. Under prodding from animal rightists and due to economic maneuvering, a number of airlines refused to handle live shipments as a part of their mail handling contracts. Hatcheries and poultry groups became organized to form a shippers' group and brought about a Congressional mandate requiring that live shipments continue as a part of the services provided by the U.S. Mail.

Still, all is not back where it once was. As a result, some airlines did opt out of mail service, several shipment restrictions appeared, there is often much confusion about these regulations, some regions are not served by air carriers that will handle live shipments, others will not handle lives at certain seasons, and shipping costs have gone up quite substantially. There are assurances in place that the mail will continue to handle live shipments, but many fear that there is no longer sufficient political clout among those who are supporting such shipments. Costs have already gone up markedly to ship live birds of all ages and the broiler/confined layer sector uses its own trucks to transfer and ship its production and young birds for restocking.

For a brief time one recent fall, the only way to access poultry genetics nationally was through the purchase of hatching eggs. Some individual producers would never sell and ship anything but hatching eggs and many still fear that the time of no mail shipments of lives will come again. The statement often heard now is that if there are poultry genetics you have ever wished to acquire, the time to get them bought and shipped in is now. I try never to be that pessimistic, but do believe that the trade in hatching eggs is on the rise again and will only gain in importance.

The Internet now is home to a great many hatching egg auctions and marketing sites. It has certainly revived interest in this purchasing option and in the early stages of revived interest in them some rare breed genetics are only available in hatching eggs. Such eggs are often offered in very small lots (three or four), are offered in mixed lots with eggs of other breeds, and sell for some quite high prices. I have seen eggs of some of the rarest of the rare offered for as much as ten dollars per egg plus shipping costs. Also, bear in mind that some states have regulations on shipping hatching eggs that are every bit as restrictive as the rules and limits on shipments of live birds. Nor is it currently legal to have hatching eggs simply mailed in from abroad.

There is not a lot of difference between hatching egg and baby chick prices when packaging and shipping costs are factored into the price equation. A friend charges nine to fifteen dollars for her rare breeds eggs and sometimes quite a bit more plus a five to nine dollar a dozen charge for packing and shipping. She literally sells the entire surplus that she cares to in the always busy, traditional spring hatching season. Getting a hatching egg from point A to point B is a challenge worthy of the best design engineers; in fact, one major university in the nearby St. Louis area each spring has an egg-based design competition for its engineering students. They are to create containment for a raw egg that will enable it to survive a one-story (ten feet) drop on to a hard surface. A very, very few eggs survive this challenge despite some most elaborate container designs. The recommended hatching egg shipping procedure is to begin with a heavy weight shipping box that will allow at least four inches of padded space all the way around each smaller container of eggs to be shipped. Into the larger carton begin by laying down a four-inch layer of styrofoam peanuts or balled sheets of newspaper. The smaller, internal containers are regular one-dozen egg cartons. The eggs go into these cartons small end down, but before seating it, wrap each egg in one sheet of paper toweling. Draw the edges of the towel up over the egg and twist them closed to form a small pouch. It holds each egg more securely in place, provides a bit of padding, and should keep other eggs in the carton cleaner if one does break. Some will use a square of bubble wrap rather than paper toweling, but it is not absorbent.

Some will fill all twelve segments of the egg carton and others will alternate or stagger the position of the eggs in the carton placing just six eggs in each carton. They will then close the carton, secure it, wrap it in bubble wrap and secure it again. Place one or two cartons on the base packing layer and pack all way around them with at least four inches of the packing material. With a deep carton, lay down a four to six-inch layer of packing material below the first eggs and place one or two more dozen on top of this. Then pack around them — up the sides and top with at least four inches of your packing material. It is probably best to try to ship no more than four-dozen eggs per shipping carton with this method.

Some believe that marking the shipping carton "fragile" or "hatching eggs" just seems to invite rough handling. I would hate to think that people could really be so callous. Our local postal workers have always been considerate of egg and chick shipments. Becoming available now are a number of new shipping containers and foam packing materials that should do much to facilitate hatching egg shipments. Much discussed is the need or advisability of insuring hatching egg shipments and to what level. Some have reported that postal authorities will only award table egg prices should hatching egg shipments be damaged or lost. An old Cochin breeder from Central Missouri once described his way around this dilemma to me. He would buy a cheap, stemmed wine glass to place in each shipping carton, label the carton "glassware," and insure it for a few hundred dollars. The insurance costs were modest and should damage occur, a broken glass could be shown. His basic premise was that the insurance papers all but assured that the carton would receive special care in transit.

At first glance, hatching eggs would seem the least-cost option for making a start with chickens, but they do not fare well in transit even with the best of packing. Handling, jarring of the egg air cell, and time in transit all combine to reduce hatchability. With shipped eggs, getting thirty to forty percent hatched has to be considered quite successful.

Some of the rarest breeds and emerging hot varieties are often only available as hatching eggs. Also, hatching eggs are the shipping option with which a number of the smaller breeders feel most comfortable. They will sell only hatching eggs and only after they have introduced enough hatching eggs for their own needs. Thus such eggs are often offered in what might be termed the off-season for hatching egg production. At the height of interest in the Penedesenca breed you would often see their eggs offered in lots as small as just three or four eggs and in assortments with eggs of other breeds. This is certainly not the best way to make a start. The numbers are just too small to create viable populations, but sometimes it is the only game available.

There are a number of steps to consider for bettering your chances of success with purchased hatching eggs.

• Have your incubator up and running in good working order well in advance of any egg delivery date. You will need a minimum of seventy-two hours of optimum running to assure that the incubator is operating at the correct temperature settings.

• Request that you be sent eggs that have been laid for no longer than three days. Yes, eggs up to two-weeks-old will hatch if stored correctly, but after about seven days, the hatching percentages easily start to decline. An older egg age and lengthy shipping time combined are just too big a deterrent to a good hatch.

• Do not order eggs to be sent when weather extremes are apt to occur. If at all possible, drive to pick up any hatching eggs or try to work with fellow producers who are traveling to transport eggs back to you.

• Notify your post office if you are expecting a hatching egg shipment, ask them to call you when they arrive, and open the container at the post office in case some sort of damage in transit has occurred and a claim for loss is to be filed.

• At home unpack the eggs, remove any broken or cracked eggs, candle them for fine cracks, and then let them stand for twenty-four hours to allow the air cell to reposition if it has been moved out of position.

• Do not attempt to incubate eggs stained with the yolk and white of broken eggs. Wipe them with a soft cloth. Scrape lightly or even buff gently with fine-grained sandpaper if the egg material has dried.

I know it is a most controversial practice, but I have had success with dipping hatching eggs needing cleaning into a bowl of warm water, about 103 degrees F. I may even add a bit of household chlorine bleach to it. Dip them for no more than thirty seconds and wipe them carefully with a soft cloth.

Use light pencil strokes to denote their origin and other essential data on the eggshells. Also note such data in your incubator operation log.

There are certainly no sexing guarantees with hatching eggs and the safe assumption will be that of every ten chicks hatched, six will be little cockerels. To build workable numbers you should order hatching eggs in at least two dozen lots whenever possible and figure on having to order multiple lots in the early going. I remember one eighteen-egg shipment that we received and that yielded just one chick. Also, with small lots there is always the risk that numbers will be skewed heavily to chicks of one sex or the other.

The hatching egg was once the most common way to market and distribute poultry genetics. A setting of fifteen eggs, enough to be covered by a fair sized hen, was a fairly common unit of rural measure. Today, for ease of shipping and handling, most hatching eggs are sold by the dozen. Even with great care hatching eggs are far from a certain proposition. The key is to buy the freshest eggs possible. Swallow hard and pay for the best and fastest shipping option available to you. Be realistic with your expectations and accept that this is probably the longest route to flock establishment. Still, they may be your only option for accessing certain, needed genetics.

If you opt to start your heirloom flock with day-old baby chicks, you will have to allow time for them to develop and purchase them in the numbers that will allow for full and proper culling as they grow and develop.

The old rule of thumb has been to order at least thirty percent more chicks than you ultimately wish to place in the breeding flock or laying house. Thus, you can be comfortable in culling the expected three to five percent of the birds at each point in their development prior to their entering the flock.

Baby chicks have been the doorways into poultry production for as long as I can remember and for a great many years before that. Many years ago I spied an ad in the back of a farming magazine offering one hundred baby chicks plus a free surprise gift for just $2.95.

I squirreled away nickels and dimes for weeks and sent off my order late one spring morning. A few days later my folks got a call from the local post office about Kelly's just arrived purchase. My one-hundred chicks

WHAT IS A HERITAGE BREED?

This corner has long championed the importance of purebred poultry to the family farm producer. They and they alone give independent flocks the genetic strengths and predictability that can be managed at the grassroots level to assure flock and farm adaptability and sustainability.

Over the past few years a lot of printers' ink was used to exhort the keeping of "heirloom" and "heritage" poultry breeds and the terms were often used interchangeably. Raised in the Midwest, heritage breeds meant, to me, the large fowl chicken breeds that were once common to the family farms of the heartland. These included not only the Rocks, Reds, Wyandottes, and Buff Orpingtons, but many white egg breeds such as the White and Light Brown Leghorns and Anconas.

The American Livestock Breeds Conservancy recently sought to set down a most exacting definition for "heritage" breed chickens. By their definition a "heritage chicken" is one "hatched from a heritage egg sired by an American Poultry Association Standard breed established prior to the mid-20th century, is slow growing, naturally mated and with a long productive outdoor life."

A "heritage egg" can then only be produced from one of the pure breeds meeting the above definition.

There is much merit in this definition although it does not include breeds like the Maran that meet the timeline and are recognized by poultry groups abroad, but not currently by the American Poultry Association (APA). Nor a breed like the Ameraucauna that did not receive its APA sanction until the 1970s. It may also raise questions about new colors or patterns to be added to existing breeds.

The American Poultry Association itself has very exacting criteria for recognizing new colors and varieties of large fowl breeds. And a part of poultry breed preservation work is restoring old colors, patterns, and breeds that have fallen by the way. I do feel confident that should any issues arise along these lines they will be resolved for the good of the breeds in question.

Two points in the extended definition that I view as very important are that the breeds must be "reproduced and genetically maintained" through natural mating and to reach a market weight at no less than 16 weeks of age. The natural matings must include both parent and grandparental generations, thus enforcing the value and predictability of purebred matings.

There are many classic crossbreds that were developed from matings of purebred heritage birds and that were used widely on American farms. They were the product of fairly simple first generation crosses and easily repeatable in the countryside. The Indian River cross, made with the Delaware and New Hampshire breeds, pioneered the modern broiler trade.

This measure won't end practices like the complex and extreme hybridization practices used by some broiler and commercial layer producers or the practice of artificial insemination used by those pursuing some rather extreme aspects of type and conformation. However, it may become a touchstone in the marketplace for those consumers concerned with the integrity and sustainability of the modern food supply system. Chickens marketed under the American Livestock Breeds Conservancy imprimatur must include the variety and breed name on the packaging.

The slow growth rate guideline recognizes not any particular shortcoming of the "heritage" breeds, but rather a long established truth that today's "fast broilers" do not look, cook or taste like the meat birds of an earlier day.

The 16 weeks of growth assure adequate frame growth, the time to allow a natural pattern of muscle development, the size and age to range and forage more efficiently, and these few extra weeks produce poultry meat of better flavor and texture. Whether New Hampshire, Delaware, Rock or Wyandotte, purebred birds can fairly and efficiently compete with non-purebreds under these parameters. One has the feeling that the Tyson's of this world aren't going to be happy until they are growing chicken breast tissue on petri dishes and harvesting them like heads of cauliflower.

Colony-type laying houses filled with thousands of brown egg laying hybrids are producing "cage-free" brown eggs for many retail market outlets. And the Cornish X broilers that can survive there can be and are marketed as broilers.

With these definitions there are white-shelled eggs that can be more profitably marketed as "heritage" production, and colored-breed broiler chickens, farm

> bred, and hatched purebred broilers that should also have greater value.
>
> Like most other family farmers I bridle at rules and regulations, but these defining efforts should do much for poultrymen in this age of the aware and responsibility driven consumers.

were Leghorn cockerels of just about every hue in the rainbow and my "free surprise gift" was two hundred more chicks just like them!

We enclosed a stall in the barn with chicken wire and set about raising the chicken flock from Perdition. You couldn't kill those rascals with a shovel and they moved about the old homeplace like so many feathered locusts. The last few we harvested were shot down out of the trees and my mail privileges were severely regulated for many months thereafter.

A good baby chick now may cost you two to three dollars and some of the rarer ones from private sources are twice that and more. Also, if they are shipped from any distance, mailing and handling can add another one dollar per chick. One of the big five hatcheries offers its very top end baby chicks for one hundred twenty-five dollars per box of twenty-five and that box may contain no more than five or ten chicks of the rarest varieties.

A number of caveats often go with baby chick purchases and especially those bought in very small lots. From most hatcheries you should assume that all chicks of the same breed are either full or half-siblings. The minor breeds are also more likely to be sold as hatched, that is they are unsexed. Most private breeders sell their chicks on a strictly as hatched basis and often can supply no more that five or six of certain breeds per shipment. As hatched chicks typically hatch in a ratio of six cockerels to every four pullet chicks. Thus in very small lots you may receive all of one sex or insufficient numbers of one gender to create any sort of breeding group. Small lots are a way to add to a flock, but seldom will one or even two small lot shipments produce enough birds from which to successfully launch a new flock.

From a shipping lot of twenty-five as hatched chicks of the same breed perhaps the best that can be hoped for is one good breeding trio and one or two additional breeding pairs. Those extra males can prove invaluable in the event of unexpected loss because of death. Also, a couple of extra males should always be held back as insurance for the next breeding season.

It is probably best to begin with the staggered purchases of two or three lots of twenty-five from different sources or a single line, whichever is your flock raising strategy. Chicks were once most commonly sold in multiples of one hundred and I can remember rural post offices filled on Tuesday and Wednesday mornings with several hundred if not a couple of thousand peeping baby chicks.

Twenty-five chick lots now seem to naturally fit today's trend toward smaller poultry flocks. They can be raised in small, inexpensive facilities and are not a bank account busting venture to start. For a great many days a set of chicks this small can be held in a brooder made from a large plastic storage box.

Order chicks as early in the year as possible. Late-hatched chicks tend to grow more slowly as the hours of available daylight decrease. They may not develop as fully and will reach productivity later in the year. In fact some bantam breeds were developed in part from late-hatched chicks of their standard sized counterparts. To get a fair number of the very rarest you may have to order as many as four or five small lot groups in a season and tap into more than one source.

Most hatcheries and breeders will ship no fewer than twenty-five chicks at a time (some will send fifteen in very warm weather) to assure comfort and warmth in transit. That means that you must sometimes contend with what have come to be called "filler" chicks. Often from the larger sources of supply they are little more than throw-ins, cost very little, and can range from sex-link cockerels to surplus chicks from orders for more mainstream varieties. Whenever possible pay the price to get something better and more useful than surplus sex-link or Leghorn cockerels.

From individual breeders and smaller hatcheries this can be a bit more of a challenge. They do not have the surplus cockerel chicks to fill in with and their other breeds may be every bit as rare and pricey as the ones in your primary order. There are a fair number of larger breeders who can ship in great variety although quite often from very small mating groups for certain breeds. Look upon these situations as an opportunity to check out a second rare or heirloom breed or to build a second flock of a traditional standard breed.

At today's prices, when shopping for baby chicks, don't be afraid to ask the hard questions. Especially with the minor breeds — you have to know as much about their breeding and background as is possible. Steer clear of any birds tabbed "not for show" or "not for project work."

After the chicks are located and the order placed, it is time to get ready for their arrival.

Have your brooder set up and warming at least twenty-four hours before the chicks are due to arrive.

Notify the post office that you are expecting a chick shipment, ask to be called as soon as they arrive, go directly to pick them up, and examine them carefully at the post office in case any losses have occurred in transit.

You could possibly refuse to accept a really damaged box or one that has been too long in transit or you may need postal documentation to file a loss claim with the shipper. One spring a friend ordered some very rare and high dollar Maran chicks that were to arrive on a Tuesday following a Sunday hatch date. They didn't arrive until the following Saturday, many were dead with the remainder in rough shape, and many of them subsequently died. The post office had obviously dropped the ball, but the shipper, a private breeder and a good guy, took it upon himself to make up the loss.

At home take the time to handle and observe each chick closely and note any obvious defects or problems.

Cover the floor of the brooder with paper toweling. It is a non-slip surface that will prevent problems with spraddle legs developing. It can be lifted in a few days when the chicks have grown stronger. Have ready chick-sized waterers with drown proof lips and a flat tray or even a cardboard box top upon which to offer the chicks their first few feedings. I like to use the plastic waterers with the red bases and feeders to help draw the chicks to them.

Warm the drinking water a bit the first time or two, and to each quart add two or three tablespoons of white sugar. Offer this for a day or two to give the newly arrived chicks an extra energy boost. Then, for a couple of days each week, add a vitamin/electrolyte product to the drinking water going into the brooder. Once a week add a capful of hydrogen peroxide to each waterer in the brooder and once each week wash the waterers in a mild bleach and water solution.

As the chicks are placed in the brooder, dip their beaks into the water to get them used to drinking from them. Old hands will also even wipe their still damp bills in the chick starter.

We provide supplemental heat with an electric heat lamp suspended twelve to eighteen inches above the chicks. We generally use 125-watt bulbs and raise them as the chicks grow during what is termed the "hardening" process. A red-tinted bulb will lessen problems with feather picking.

With bantam chicks, birds closely inbred or others needing a quick boost use the more nutrient dense and smaller particulate game bird starter instead of regular chick starter.

In an emergency situation where you have no starter on hand offer the chicks what they have already been feeding upon — egg. Finely chopped, hard boiled egg at room temperature can be offered to them at floor level for a few minutes several times each day. Offer only what they will eat in a short time and then carefully clean up all the remaining traces.

Shortly before hatching, the chick absorbs the egg yolk into its body through the umbilicus. This gives the chick enough nourishment to sustain itself for up to 96 hours following hatching. Thus the day-old chick has the needed wherewithal for 24 to 96 hours in transit through the mail.

For most folks, baby chicks will be the starting point for their poultry venture — be it rare, heirloom or traditional poultry breeds. Baby chicks are now the means that are accessible to the greatest number of producers. From a box of as few as fifteen chicks, a start can be made and I can think of no other purebred livestock venture with lower beginning costs.

Adult or started birds will give the beginning producer the fastest start. It is also the start that can be made with the greatest assurance of quality as you are beginning with made birds and in the desired ratio of the sexes. It is also the most costly way to make that start. Their worth is determined by the quality they manifest and their producer has a substantial investment in bringing them to a more developed stage. With adult birds you can be into production in literally a matter of weeks and with birds over which you have had great control in selecting. Factored into the costs of this option must be shipping fees, too. These can be very nearly as great as the cost of the birds themselves.

There are, roughly, two seasons of the year when adult birds are most readily available. The first is late-May to late-July when a lot of hatcheries and higher volume producers are dispersing their breeding flocks for the year. These are birds with a lot of lay left in them, but are no longer needed because the chick-shipping season has largely come to an end. A few of these adult birds will continue to hatch into the early fall and a very few hatch the year around.

The second adult bird season is in the late fall after all breeders have broken up their breeding pens and are making the final culling of the birds of the year. At this time you may access older birds that are being replaced by younger ones or the adult birds may be the last to be culled from birds that were being grown out as replacements.

These are generally good birds, but just not the elite of the year. You cannot expect another breeder to let go of his or her very best. However, much can be done with birds that are slightly past their prime or are the sibs and cousins to an established breeder's better birds. Birds bought in the summer can actually produce a goodly number of chicks in the weeks that follow.

Such birds should he transported gently, handled with no break in feedstuffs variety, and be kept comfortable and well supplemented through any hot weather stress. A few years ago we traded some Buff Wyandottes for Blue Wyandottes from a breeder in Ohio. They were birds coming straight away from a show there. He placed them in solid-sided transport boxes, loaded them in his van in the evening, and pulled into my front yard at daybreak the following morning. One hen laid the second day, both were laying steadily within the week, and we had chicks less than a month later.

A good breeding age bird now will cost from ten to one hundred and fifty dollars. Yes, this is quite a range in price, but there is also quite a range in birds now available. Factors like rarity and quality enter into the pricing along with the cooperative nature of some producers anxious to see their chosen breed placed into more hands. The first pair of Penedesencas that I was aware of being offered for public sale brought a bit over seven hundred dollars at auction. They were a young pair of around eight weeks of age. Two years later similar pairs were selling for twenty dollars at our local farmers' market.

Adult trio prices normally fall into the thirty-five to one hundred dollar range. I once paid thirty-five dollars for a young pair of birds I thought I absolutely had to have and have sold a couple of seventy-five dollar trios. If sending such birds through the U.S. Mail, expect to pay ten to twenty dollars per bird for postage and the proper shipping container.

The U.S. Mail will accept live, adult birds if shipped in containers with the approved liners designed to check microbe spread. The boxes with replaceable liners can be reused with fresh liners. The boxes are sent as high priority mail and should be sent no farther than distances that can be covered in forty-eight hours or less. Into the boxes will be placed a couple of apple halves or some of the new hydrating gel products to keep the birds content and hydrated while in transit.

With the purchase of adult birds you gain at least a full year of production time. Also, you can acquire birds of proven conformation and color and in the exact numbers desired.

Ideally, you should acquire pairs or trios from two or three distinct genetic sources from which to then begin assembling your own breeding line. If finances won't stretch that for — mine seldom do — try to acquire a trio of not too closely related birds from a single source. You don't want to begin with a complete outcross, but rather from a base that will support heavy breeding pressure while maintaining a fairly high degree of genetic consistency. The good ones will be truly prepotent for the traits that really matter.

Keep the new arrivals isolated from other birds on the farm for at least thirty days after their arrival. During that time monitor them carefully for health matters while really studying their type and breed character. Tend them last each time you do chores as an additional health control measure. Try to continue the new arrivals on the same ration they were receiving prior to being moved and enrich their drinking water with a good vitamin/electrolyte product. To smooth their transition, keep them dry, protected from drafts, well fed and watered, and in a unit that provides them with both adequate headroom and floor space.

There are several steps to acquiring the best adult birds possible for your needs.

Begin by preparing a list of potential breeders to be contacted. Such names should be available in periodicals like the *Poultry Press* and *Acres U.S.A.*, the state and national directories for the Poultry Improvement Plan, and the American Poultry Association and breed groups. Much help is also available from the American Livestock Breeds Conservancy and the Society for the Preservation of Poultry Antiquities.

Contact prospective buyers with an early evening phone call or by mail with a letter including a return self-addressed stamped envelope.

Be very clear as to your wants and realistic in your price expectations.

Give yourself plenty of time for this shopping process.

Always seek quality over quantity, even if it means starting with as few birds as just one good pair.

As a group, poultry producers are good people and I have had only one bad experience with ordered birds from a private source over the years. Generally, when filling an order they have nearly all followed the old timers' policy of "heaping the measure." I once scrimped and saved to buy a trio of Rhode Island Red bantams when adult birds had to be sent by air freight only from one large airport to another. When the shipping box was opened, there was my good breeding trio exactly as promised and another good pair, a gift from the breeder.

Don't count on there being a great distance between sources to assure that you are getting birds or eggs of different backgrounds. You must ask questions of even the largest of hatcheries. If they can't or won't answer your questions then they don't need your money.

Neither live birds nor hatching eggs should be a spur of the moment purchase. Order them for a convenient delivery date, be fully prepared for their arrival, and communicate often with the supplier as questions arise.

Earlier in this book I touched a bit on the subject of seedstock costs, but I should go a bit deeper into this matter. I could throw around a lot of old clichés such as "the good ones don't cost — they pay," but what does this really say? Not everyone needs a three hundred dollar trio of breeding birds. Their twenty-five dollar siblings that won't win at major shows should have the genetic potential to produce offspring every bit as good. Remember, when sound purebreds of the same variety are mated together they should produce at least to the average of the breed.

Interestingly, poultry ownership has long been linked to financial well-being. In the classic musical Fiddler on the Roof, the character Tevye sings that one of the things he would have if he were a rich man was a yard full of chickens and other poultry. Here in America, "a chicken in every pot" and "chicken on Sundays" were both metaphors for financial well-being and security. To this very day, poultry breeds are kept by royalty and it was Queen Victoria's Cochins that launched modern poultry breeding, as we know it.

I would not be so bold as to say what anyone should or shouldn't pay for a bird or birds, but believe that today a better case can be made for buying a one hundred dollar rooster than for a five thousand dollar bull. It is hard to put a cash value on a bird that will move your breeding program forward even a few strides. Simple economics have long favored poultry keeping on a modest scale. A hen begins to lay at a relatively young age and a meat bird can be harvested within a very few weeks of hatching. The hen produces eggs, chicks, her own replacements, and when no longer productive, can be tuned into a tasty pot of dumplings or soup. Ten dollars for a single rare breed hatching egg can seem to be a most daunting purchase, but few are the livestock species that can produce themselves as quickly and in such numbers as the chicken can.

Purebred poultry production is now poised to take a big step upward into a position now seen with other types of purebred livestock. They could return to the position that they held in the thirties and forties. As breeders become established they should be able to go on marketing birds of exceptional merit and value.

In the early going with the rare and heritage breeds you will be in what essentially can be called a "numbers building stage." You will be working to put together sufficient numbers to breed from and support sufficient culling for a true quality-building breeding program.

Thus, at the start, every effort should be made to begin with the best possible birds. "Best possible" is the key phrase here.

When a scant handful of a breed remains or only a few have entered the country, every bird counts and you will have to draw from what nearly every bird has to offer. The preservationist and working flock builder is not about winning in the showroom and may have to leave the genetic fine tuning of some traits for a later day and later generations. A crucial bird today may become quite expendable a year or two down the road.

With such small numbers to work with, tolerance is needed for some modest defects or shortcomings that have a lesser impact. Try not to breed in any glaring faults, but neither should any vigorous and hardy individuals be cast away just because of a few simple flaws. Breed these birds to others that are exceptionally strong in the areas where they are the weakest.

Over the years we have acquired many breeding trios in which one female was clearly better than the other. Generally, in the second season we would have numbers sufficient to enable us to let the weaker female go. For that first year, however, she was certainly good insurance. In some instances that hen may have actually nicked better with that particular male than her better appearing pen mate.

If both females were siblings they would each carry the genotype and could be expected to produce comparable offspring. With modest numbers always move cautiously and don't let go of any bird until there are two or more in hand that are better and more proven birds.

In this early, numbers-building stage, chicks will be raised in good numbers, new bloodlines will be trialed, and birds will be added to both upgrade and further invigorate the line. Most good flocks will over time, evolve out of one or two good matings. The flock could ultimately trace back largely to one exceptional individual, very often a female. A flock can be maintained with frequent genetic inputs from various outside sources, but it will lack predictability in its offspring and the long-range production pattern.

In commercial poultry production there is a long-term tradition of turning the flocks often and "in toto" meaning completely. Broiler chicks and pullets would be bought off farm. Broiler chicks would be turned perhaps two or more times a year and laying flocks would be dispersed at the end of but one laying season. This short duration production has become a part of the economic cycle for commercial poultry production, but it has left some with an almost throwaway or highly devalued view of chickens.

A chicken, given good care, can remain quite productive for a number of years. Following the first or second year of lay the number of eggs a hen will produce in each cycle will decline. Still, if she is producing offspring as good or better than herself there's no reason not to consider continuing to breed from her. A hen may remain reasonably productive for as long as ten years and that kind of durability should be cultivated in any breeding program.

Likewise, males have a long productive life if they remain sound and do not become overly large and ponderous. Many will hold and breed from their better birds until they are somewhere around four years of age with only modest declines in production from year to year.

There are a lot of bits of folklore and old wives' tales about bird ages. Some say that breeding older males to pullets will increase the number of female offspring or breeding young males to older females will do the reverse.

A number of veteran producers will not keep back offspring from any birds until they are at least in their second year of production. There are some sound reasons for this. The first being that such birds are at this point now starting to develop a track record and their vigor and the quality of their performance have become more clearly discernible. Also, older birds in the flock that have demonstrated an ability to live and thrive in that environment may even be imparting some degree of natural immunity to the particular "bug" soup faced by their offspring.

The real influence of a good male may not be felt completely until his daughters and granddaughters have gone into production in the breeding flock. It is commonly held that in each mating it is the female that most influences type and that the male most influences color and pattern.

In your heritage breed research you may read often of "male matings" and "female matings," sometimes they will say "lines" instead of "matings." This is generally a practice of exhibition breeders of some of the varieties with more elaborate coloring and patterns. It is a belief that some breeding lines within that breed or variety produce better males and others better females. Deliberate breeding selection is done for color traits that will produce better representatives of one sex or the other.

This is a practice that appears to be totally at odds with the real world practicality that most aspire to with a sustainable farm flock. If it is truly needed it raises a most serious question about how closely genotype and phenotype heredity traits actually are in such breeds or how valid and practical some patterns really are? I am taking a quite contrarian position here, but if you are building your purebred flock based on practical roots you should not need to maintain such segregated breeding lines.

There is a long held observation that poultry producers break into two rather distinct groups as to numbers and output. If a breeder in each group is given the task of producing thirty good chicks of a certain variety for the year, one will raise just thirty and the other three hundred. This illustration is perhaps too simple and I'm not completely sure that it can be construed as a valid criticism of those that fall into the latter group.

There are some flocks out there that are as tight and fine as jeweled watches. They are the work of a lifetime, sometimes multiple lifetimes, and no expense has been spared in putting them together. Others have not reached that point yet and they may never do so, but they can still produce some very good birds.

A poultry flock is very reflective of the person or family that are building and managing it. No two will ever be exactly alike and they may be operated for any number of different reasons. During the course of writing this book I had to take one of my flocks completely apart when a genetic defect began appearing. Our poultry yard never will win any prizes and I am a very poor carpenter and design engineer. Yet, I have learned one lesson that is paramount about such things. A breed building cock bird will drink just as contentedly from an old tomato can wired into the corner of a cage as from a fifty-dollar waterer.

The birds are what make the flock and if you are working to make next year's crop of chicks better than last year's — then you are doing a good job.

As you set about selecting your breed or breeds of choice, the question before you will always be, does the world really need another white or black or even blue chicken? Probably not for most of the basic reasons, but life teaches that what appeals to the eye and the heart ultimately comes to appeal to the mind.

To be truly successful with any pursuit in life you honestly have to like what you're doing and truly value that with which you are working. Trace them back and virtually every chicken breed was developed for a practical purpose.

Over time things like color and extremes in size or feathering may have been a bit overplayed, but I know of no breed that could or should be dismissed totally out of hand. A few years back I was at a market and saw a pair of Buff Laced Beardless Polish birds that were virtually without flaw. They were of such size and good layer type that they would have caught the eye of any range or cage-free egg producer. They were indeed something to build upon and while no one can keep all of the good ones that come their way I really wish I had bought that pair.

Breed choice does depend upon the task intended for the flock. This includes thinking at least ten to twenty years down the road every time you ponder taking up a rare or heirloom breed. Are birds needed for a fancy table egg flock, to produce meat birds or to sell replacements and breeding stock to other producers? There are heirloom breeds for every task and each one can even be made better with simple, selective breeding. The local egg market won't be sewn up any time soon with a flock of Blue Orpingtons. Nor will Don Tyson of Tyson Foods be sent a shiver down his spine with even a three hundred-hen flock of White Giants. The goal at this point should be to take the first steps toward serving the market at hand with the kinds of birds and production methods most favored by your customers.

Producers are unlikely to sell a great many Speckled Sussex or Russian Orloff chicks even in a very good year. However, they can begin assembling a modest flock of such a breed that will produce first-rate chicks — a true source flock. The producer with a pure commercial bent, be it eggs or poultry meat, will seek to build a flock of a single breed strong for that particular trait. Some of the rarer breeds are just not that productive or their productive traits have been long neglected in various breeding programs. For example, there are hens out there that will peak at ninety eggs per year, are quite seasonal in their laying patterns or that will have very high maintenance costs due to their size and slow development.

For some needs you may be best served by just a ten or even a five-bird flock and for other options you may need to build to a flock of three to five hundred birds. Such numbers are doable without skewing a small farm into a "too specialized" direction. Ten heavy breed hens at eighty percent lay and with eighty percent hatchability should produce forty-five meat-type chicks a week during the peak hatching months of February through June. A good source of stocking for a range broiler operation thus will take up very little space and be far more dependable than hatchery chicks that may or may not face future shipping challenges.

A source or seedstock flock need not be massive either. Numbers commonly given are twenty to forty females and five to eight breeding males. Such a flock can be operated in multiple larger breeding pens and with several, smaller side matings.

A purebred laying flock is truly self-perpetuating. It may even produce chicks for sale over and above those needed to replenish the existing flock. Again, ten select or elite hens pulled from the larger producing flock, could easily produce at the levels listed above. Thus they could supply twenty to twenty-five replacement pullets per week for many months of the year.

Rare and heirloom breeds with a history of use for poultry meat production would include the Buckeye, Chantecler, Cornish, Dorking, Java, Giant, Naked Neck, Orpington, Plymouth Rock, Delaware, New Hampshire, and Wyandotte. None, at this moment carry a level of performance anywhere comparable to that of the Cornish X broiler lines, but these lines were developed from two of the pure breeds noted above.

For the United States meat bird trade, the clear preference is for birds with yellow skin and legs. White feathering generally means an easier bird to dress. This need not rule out rearing colored varieties and breeds with white skin, but these are factors that must then be addressed in the marketing process.

The Cornish X broiler represents a good half-century of agro-genetic engineering and is probably something independent producers wouldn't want to duplicate even if they could. The goal here should be to selectively breed toward a twelve-weeks-to-harvest purebred broiler that looks like a real chicken, lives like a chicken should, and tastes like a chicken. Would our grandparents now know what to make of what is today called a broiler? If not, what business do we have trying to grow it and sell it?

Along with the yellow skin, American consumers clearly favor a broiler with a three to four pound dressed weight. Nor are they really comfortable with birds that are marketed as the whole bird.

One reason white feathered birds are touted for their dressing quality is simply that white pin feathers are less noticeable. Nor do you see any of those inky stains in the skin from dark feathers. Birds with desired feathering and skin color would then include White Rocks, White Wyandottes, and White Cornish. The latter, while heavily muscled, are quite slow growing. The Delawares and varieties with the (largely white) Columbian color pattern and yellow skin would also be good choices. Although a bit smaller, Rhode Island Whites would also be worth considering.

Into a second group would go those breeds with both growth and yield potential. Breeds here would include Barred and Buff Rocks, Rhode Island and New Hampshire Reds, Dark Cornish, Dominiques, and a new breed, the Braggs Mountain Buff. The latter was bred to be a fair layer of large brown eggs and a meat bird comparable to those being produced for the Label Rouge broiler program in France. The Dominique is a very old breed with a distinctive flavor that many compare to the taste of pheasant.

A third category would be meat birds best suited for the rather specific, meat bird niche markets. For example, certain ethnic markets favor black males of some size such as Black Giants or Javas. A second choice for some of these markets may or may not be red-feathered birds. Some ethnic markets will not accept white-feathered birds. Also, some will reluctantly accept a few females in lieu of males.

A few years ago specimens of some of the heritage breeds were sent to a select group of West Coast chefs for cooking quality and taste testing. Superior, easily detectable differences were noted between these birds and those available to them through their regular supply outlets. It is reasonable to believe that, just as beef and pork from different breeds have different eating qualities, so too should poultry meat from different breeds.

A breeding flock kept to produce broiler chicks on the family farm does not have to be huge. Twenty hens at eighty percent production and eggs with eighty percent hatchability would produce eighty-five broiler chicks per week throughout the spring and early-summer grow-out season for broilers. Key here would be to keep sufficient cocks with the hens to maintain fertility without having the hens be worn ragged. With extra heavy birds I would drop to one male for every four to six hens.

Closely monitor the hens for wear and excessive feather damage. Also with too many roosters, matings may be interrupted and incomplete. We once supplied White Rock eggs to a Missouri hatchery and they wanted one breeding male kept for every four to five hens in the flock. With this stocking rate, the hatchery regularly got hatching rates of ninety to ninety-five percent. By the end of the hatching season in early June, the hens had substantial feather loss and weren't pretty, but were still heavy and laying well.

Due to their size and tendencies to become more ponderous with age, it might be wise to turn this breeding flock every year or every other year. These can be replaced with the select from their offspring. Thus some off-season costs can be pared and there can be assurance of more vigorous and agile breeders.

To my mind, the purebred flock producing broiler stock will need even more genetic tweaking and upgrading than most laying flocks. The meat-producing role of many of these breeds has been all but completely neglected for decades. Many of these large breeds now take many months to simply develop frame before putting on any muscle mass.

To say the unthinkable, the quest now should be for a smaller Giant, a wider Rock, and a denser Wyandotte. For the real experimenter, it is time to dust off those Buckeye and Chantecler genetics. Down the road, I believe the more different your birds can appear from the Cornish X while still performing well, the better.

All chicken breeds lay at least a few eggs each year or they would cease to exist. Some just do not lay a great many eggs, follow a very seasonal laying pattern, lay small eggs or are of a very high-strung, excitable nature. For example, it used to be said of some Cochin lines that the hens would lay a dozen and a half eggs in the month of May whether you wanted them or not. The other eleven months of the year were pretty much on your own. With the rare and heirloom breeds your egg layer options are truly many and varied. Breeds can be selected with shell colors ranging from blue, green, chalk white, parchment, light and medium brown, and the deep brown of a chocolate bar. There are breeds that will consistently lay 160 to 180 eggs per year in simple housing, while some will be hard-pressed to lay ten dozen a year, and some with the potential to top two hundred eggs per hen per year. Nearly every one of the breads could be bred up for better egg production and even the heaviest breeds have a history of producing some quite productive layer lines.

Too many people are now fixated on the brown egg layers within the heirloom breeds. I think sometimes they expect a bit too much from certain of these breeds. A Rhode Island Red or a Rock will never lay in the same league with a performance bred Ancona or Minorca. Also, there has been a bit of genetic tampering with some birds to boost their production levels.

Beware of birds listed as "production," "performance," or with other such hyped terminology. It is a safe bet that not too far back in the ancestry of a "Performance Red" you will find at least one Rock or Leghorn rooster.

When I need to make a real "gee whiz" illustration at a farm show or seminar I will set a hatchery run "Production Red" side-by-side with a good, purebred Rhode Island Red. The colors could not be more different and the purebred bird is generally larger and more substantial in its structure. Good purebreds may take longer to build with to attain a desired level of production, but such a flock will then continue to have a long, high level of genetic predictability.

Nor is there any such thing as a "sex-linked" breed. Sex linkage is a breeding device that enables on-sight sexing of baby chicks at the moment of hatching.

For most of the history of poultry keeping the ability to sex very young birds was much wished for, but simply not possible. Vent sexing was developed in the Far East and is very accurate in the hands of an expert, but is a skill not easily mastered. Some chicks are actually lost during the sexing process and a scant handful of people do it now as a trade.

Sex-linkage takes advantage of the fact that when certain birds of different colors or patterns are mated together the pullets will hatch with one color pattern and the little cockerels in another, somewhat different one. Crossing birds that carry a gold color factor with those that carry a silver factor derives most red sex-link birds. Many of the black sex-links are derived from crossing a barred bird with a red bird.

Sex-link chicks will receive the boost that is hybrid vigor. They will generally hatch larger in size and of a bit more vigorous nature than their purebred parents. These factors will stay with them and continue to enhance their performance throughout their lifetime. Still, offspring traits like good egg-laying performance will be most directly influenced by the inherited egg-laying abilities from their purebred parents.

The real problem is that when a sex-link male is mated to females of the same breeding, a great deal of the genetic variability will emerge in the resultant offspring. Some pullets may resemble their dams (female parent) and may even be good layers, but not all. In future generations the breakdown in breeding predictability and reliability will be even more noticeable.

Purebred flocks producing strictly for the egg trade may benefit to some degree from rotating in males of dissimilar purebred breeding — not birds of a different breed, but of a different genetic background.

If they are well and carefully bred something akin to heterosis will manifest itself in the resulting offspring. Here they will be selectively bred for improved egg output, egg size, or earlier egg production. Well bred, linebred individuals should prove exceptionally prepotent when used in this manner.

Many farmers who are primarily egg producers in their focus may opt to keep a primary laying flock and a second, smaller replacement breeding flock or even several small side mating groups. On some farms there may even be separate egg and meat bird flocks.

Now available to the heirloom breed egg producer there are essentially four eggshell color choices; white, brown, dark brown, and the blue-green category. Each fits into a number of marketing niches and many will overlap. However their greatest value is, and always will be, as a product from (and representative of) the family farm on which it was produced.

Industrialized agriculture early on chose to go almost entirely with white eggshell lines developed almost entirely from hybridization within the White Leghorn breed. Thus for decades, the white egg was the only option for most urban consumers. The brown egg was seldom seen, as it was perceived to be rare and in short supply. When brown-shelled eggs were encountered, they were bought almost exclusively from small farmers. The brown egg was thus the epitome of the "fresh country egg" and was then perceived to be of greater value.

In recent days the demand for the brown egg has ballooned. Now emerging are efforts by some to produce it at the factory farm level. Brown eggs are even showing up at the big box stores where they sell for three to four times the "per dozen" price of their white shelled counterparts.

Compared to the white egg layers, brown egg layers don't handily fit laying cages, they take up substantially more floor space in loose housed systems, they eat more to maintain themselves, and their larger eggs can be a problem for mechanical handling and grading equipment. In most instances there is more hand labor with a brown egg and this further increases its costs to produce. As many brown eggs grade out as extra large or jumbo size, they require special, larger cartons also.

The letter grade on an egg refers to its visual appearance. All table eggs should be smooth shelled, uniform, and clean. Those eggs rated Grade A can be small, medium, large or jumbo in size. They can also be of different colors. Size grades are based on weight per dozen. The letter grade has nothing to say about how the egg was produced.

The typical supermarket egg today may have been in the production and transportation loop for as long as thirty days since being laid. Crack one of these and the white will run across the bottom of the pan like so much water. A fresh egg stands up on the pan and generally has a much more vividly colored yolk. Eggs, fancy table eggs if you will, now fall into a great many production categories not encoded into any sort of grading system. Production options that are added value in the marketplace now include free-range, loose housed, humanely produced, organic, additive-free, omega-3 boosted, of exceptional color, from heirloom birds or any combination of the above. However be wary of trying to pack too much into single carton of eggs.

The production options chosen to employ will dictate the laying breeds or varieties from which they should be selected. The heritage breeds offer a number of brown egg choices that will produce in the 160 to 180 eggs per hen per year range and some white egg breeds that may bump those numbers up by as many as thirty to fifty eggs per year.

In free ranging and cold housing situations, best served would be the modest sized (finer boned), clean legged, and naturally vigorous breeds and varieties. For a great many generations, many of today's brown egg breeds have been bred for little more then large size and feather quality for showroom competition situations. When propagated for laying, a lot of the brown egg breeds hold to a more streamlined appearance and mature with weights in the five and a half to seven pound range.

The traditional brown egg-laying pure breeds are the White and Barred Rock, Rhode Island Red, New Hampshire, Delaware and Australorp. Once and for all time let's kill a hoary old myth here. A great many white feathered breeds and varieties lay brown eggs. And a great many colored varieties lay white eggs. The rule is red ear lobes mean brown eggs and white ear lobes lay white eggs. That said, the Penedesenca — layer of perhaps the darkest brown of all eggs of the moment, has white ear lobes. I guess that makes it the exception that proves the rule.

Brown egg breeds from the rarer breed groups have some real potential as practical layers in simple housing include; Dominiques, Marans, Naked Necks, some Orpingtons, rarer colors and patterns of Plymouth Rocks, both combs of Rhode Island White, Welsummers, and some Wyandottes. No, these breeds aren't going to be filling Wal-Mart stores any time soon, but all are ripe for selective breeding to improve their egg output.

With all of this said, let me also add that if your heart is set upon a laying flock of Dark Brahmas or Buff Cochins then go for it. Just bear in mind that their large size, later maturity, slow growth pattern, feathered legs, and rather seasonal laying pattern will all have to be factored into the time frame and breeding process needed for them to become better layers.

The larger breeds tend always to grow more slowly, reach sexual maturity at a later age, and lay fewer eggs per laying cycle. This is said not to be disparaging of them, but rather to make everyone fully aware of the work still needed to be done with a great many of the heritage breeds. Some Buff Orpington lines of late have caught flack for reduced egg production and small egg size. Some Delawares have produced eggs that were too large. These and other heirloom breeds were once noted for a great number of efficient and highly productive breeding strains within their ranks.

A part of preservation work has to be trying to find vestiges of those strains and then to begin building upon them again. Alas much has been lost in the decades since all chickens were expected to work for a living by contributing to the family table and income. Many once valuable lines have no doubt been lost, but one of the things that drive preservation breeders is the hope that somewhere in the rural landscape a few bits of those venerable old breeding lines might still remain.

Such populations have been found before and some parts of the nation are still believed to be good hunting grounds as they have been less touched by the industrialized approach to production agriculture. Some look to parts of the Northeast and others to certain areas of the South for something quite valuable that could be living in a small pen behind a barn — maybe right next door. I have a good friend that will actually stop and knock on strange doors when he spots an interesting looking flock in his travels.

Let's take a short look at each of the egg color categories and see how they might fit on today's egg producing small farms and holdings. The white egg has been historically important to small farms since early in the twentieth century when White and Light Brown Leghorns were the feathered treasures of America's small, family farms.

Despite all of the media play that the brown egg gets, I have to believe that the time is long past due for small farmers to renew their ties with the historically significant white egg breeds. Their space needs are roughly half of those of some of the larger, brown egg-laying breeds. Whereas the brown egg breeds generally need a stocking rate of one male for every six to eight females, the smaller, more active breeds can stay fertile with a stocking rate of one male to every ten to twelve females.

While these birds do have a more active metabolism they do tend to produce a dozen eggs on substantially less feed. Their larger combs and wattles may present freeze and frostbite problems in very cold weather, but, as will be shown later, there are preventive measures for this.

The Mediterranean breed class is known for its productive layers and a great many of the breeds within it now have minor and even rare breed status. Minorcas, Anconas, colored Leghorns, even Andalusians and Spanish have a history of being a most productive layer of white eggs. These are the breeds with the capacity for two hundred plus eggs per hen in yearly production. The literature has numerous examples of Leghorn hens near or at the three hundred egg mark.

The white egg can be made everything that the brown egg is now perceived to be. It can be produced in larger sizes, humanely, organically, additive-free, cage-free, loose housed, on range, with good fertility or in any combination of the above. Their need is for more marketing and a better relating of the simple truth that all eggs are equal regardless of shell color.

I believe it is safe to say that sex-link varieties and "performance" or "production" Reds today produce the majority of the brown eggs in the marketplace. These are certainly not purebreds, far from it, and are pretty much managed like the commercial white egg layers. They are birds with a hotter metabolism that are pushed hard and are generally discarded after just one laying season. The commercial sector balks at keeping layers for more than one season due to the slightly higher added cost of carrying them through a molt and then compensating for the handling of the larger eggs that they produce in their second laying season.

Brown eggs come in every hue from a pale shade termed "parchment" through a pale brown to ever-deeper hues. A few are even laid with darker spots and flecking. Our Mottled Javas laid those pale parchment-hued eggs that didn't seem to really fit anybody's idea of egg color.

Our Wyandottes lay a paler brown egg than our Rhode Island Reds. There can be a certain amount of variability within a flock — even one made up of females that are all full and half-siblings. Old-timers tell of some Rhode Island Red lines that laid a quite dark-shelled egg. The medium to light brown egg of the Rhode Island Reds now is the shade that most associate with brown eggs.

A great number of the heritage breeds are brown egg layers and to many the brown egg is the heritage bird's trademark product. The red hen is the classic producer of said egg. Still, really good, true breeding, mahogany red Rhode Island Reds and solidly put together New Hampshire Reds are not all that abundant. A friend has a small flock of Single Comb Rhode Island Reds from a line known to have been in existence for nearly one hundred years. The hens of that line fairly consistently produce one hundred sixty eggs each per year.

A lot of the heirloom breeds now fall into the one hundred forty to one hundred fifty egg range of yearly production. Breeds from Dominiques to Orpingtons tend to fall into this grouping. Most of the Rocks are there too, although some of the White Rocks are better egg bred. If pinned down to pick just one brown-shelled laying breed my choice now would be the Single Comb Rhode Island White.

Although not currently sanctioned by the American Poultry Association, this pure white breed is a bit smaller and more active than the White Rock or Rhode Island Red. On our smallholding they have been real laying machines. Here the pullets have always begun to lay early and produce only a handful of smallish eggs before hitting their stride. The hens are generally the first to lay each breeding season and remain productive the longest.

All chickens lay eggs and through selective breeding all lines can be improved for this trait. With each passing season the flock should become at least a bit more productive. Breeding the flocks' best laying females to the sons of exceptional laying females does this.

The new, hot color in eggshells now is a very dark brown — a brown the color of chocolate or a deep brownish red. Only a relative handful of breeds are available now that produce such eggs. These breeds are the Welsummer, Maran, Barnevelder, and the Penedesenca with the Penedesenca laying the darkest shelled eggs. The color actually darkens as the breed's laying cycle continues.

These colors go onto the surface of the shell in the last stages of its trip down the oviduct. The colors are often not fully set when the eggs are freshly laid and can be smeared by nesting materials or early handling.

Some of these eggs are of a solid hue and others are dappled with darker spots up to the size of a match head.

Some maintain that the deep brown color of the Maran, Welsummer and Barnevelder eggs is the same from the first day of the laying season to the last. This has not always been our experience. Here the first few eggs have been a bit lighter and uneven in color, then they peak a deep brown color late in the laying cycle. The last few to be laid continue to be quite dark in color. This dark color needs to be continually bred for and the key seems to be to retain breeding males only from the darkest of the brown eggs laid.

The Maran and Welsummer are the most commonly seen of this group, with the Maran now being bred in several colors and patterns beyond the most commonly seen Cuckoo pattern of irregular barring. These birds are all of European origin, were developed to be true farmyard fowl, and are quite vividly colored.

The Barnevelder has a most distinctive double-laced pattern and perhaps exists here in the smallest numbers now. They are regarded as somewhat shy breeders and lacking in the vigor of the others.

The blue, green and pink-shelled eggs of the Araucana have enjoyed frequent waves of popularity. Interest in them has spun off a variety termed the Ameraucana and from them have come a rather generic, green egg layer that has come to be somewhat dismissively termed "the Easter egg chicken."

The Araucana is a rumpless fowl that was developed in South America and has facial pedicles or fleshy growths. As a rumpless fowl they have problems with successful matings and they also carry a lethal gene. What that means is that one-fourth of the baby chick embryos will develop, but then die in the shell shortly before hatching.

These are the birds that lay the vivid blue eggs and the even rarer pink ones. They are the only ones to do so consistently — as long as they continue to be bred pure. The Ameraucana is somewhat derivative of the Araucana, but has a muff instead of pedicles, has a tail, and does not carry the lethal gene. Too often they are wrongly tabbed as Araucanas and even offered for sale by that name.

The Ameraucana is now backed by a strong group of breeders and is being bred in a number of colors and feather patterns such as black and wheaten. They lay predominantly dark or olive green eggs with the occasional blue egg layer.

The Ameraucana would be the choice for those wanting to introduce eggs of a consistent color and from birds of a consistently more productive type. The key here is to only retain breeding males hatched from eggs of the desired color. This is the only way to keep blue in the line or retain a clear, lighter green.

Offered often in catalogs and sometimes dubbed "Americaunas" they are a rather motley colored group of birds that most in the know dismiss as "Easter eggers." They are splotched and splashed, tailed and often have incomplete muffs. They lack the consistency of a true breed.

Unlike the heritage and purebred flocks, the commercial laying flock is the project of but a single season and no more. The pullets come onto the farm as day-old chicks or twenty week old, ready-to-lay pullets. The breeding behind them is not fully known as it has been done elsewhere and the control of the all-important genetic material is conceded to others. The commercial birds live a life of a bit over five hundred days and more often than not, wind up in a landfill or compost pile.

A well planned and managed, purebred flock based on one of the heritage breeds will live on for generations — both in bird and human terms. It is self-renewing and should be in a constant state of upgrading and improvement. Plan to replace twenty-five to forty percent of the members of the working flock each year. Also, the flock should be slowly increased to the desired level of production.

Successful selective breeding demands a start with birds of good breed type, which are strong and vigorous in nature, and demonstrate the body type for the task. Begin with healthy and well-formed birds that are up to the task of a long and productive life on a farm or modest facilities holding.

SELECTING LAYING FLOCKS FOR THE SMALL FARM

There are a few general thoughts about laying flocks for the modern small farm that should be considered.

The sex-link birds right now will give you a great many light brown-shelled eggs of fair size, but they won't build a sustainable and enduring flock.

Small producers often need to do a better job of presenting their eggs for sale.

Even if a flock is made up of all heirloom breeds, a badly mixed up flock will not produce uniform eggs for sale, produce predictable replacements or foster a positive image. A friend says such flocks look like "Grandma's chicken yard."

An egg is an egg once the shell is removed and no one will prosper by fostering and spreading old wives' tales and misinformation. The white-shelled egg deserves the small-scale producer's consideration every bit as much as the brown-shelled variety.

A good laying flock with a purebred basis is a long-term pursuit. Don't take up heirloom birds on a whim and then neglect or let them go after a season or two. Such birds seldom make it to another set of caring hands with any sort of commitment to their preservation as a breed.

Heirloom breed producers can and should function in a number of different roles. Yet, even with a single focus, be it meat, eggs or seedstock, each flock and producer will have its own unique nature.

A part of the task is to know your breed or breeds fully and even more so the birds that make up the actual flocks. A White Wyandotte and Rosecomb White Leghorn have a great many similarities, but all must admit that they were bred and refined for two rather different tasks in life. If you have a good market for light brown eggs in fair numbers and some demand for broilers or roasters, then the White Wyandotte should be your choice of the two breeds. While Leghorn cockerels were my grandmother's favorite choice of young birds to fry, the Leghorn must be your breed of choice for eggs in greater numbers.

There is a term some hatcheries use when attempting to market birds to small holders. They have chosen to promote certain breeds as "dual-purpose" fowl. It is a term most heirloom breeders have chosen not to use.

No breed can lay like a Leghorn and yield poultry meat like a large Rock or Wyandotte. The four breeds most touted now as "dual-purpose" are the Barred and White Rocks, Single Comb Rhode Island Reds, and the New Hampshire Red. They are all good breeds and each was developed for a rather specific farm environment and performance role. Possibly the White and Barred Rocks hew as closely as any to a general-purpose role, although one does dress cleaner than the other. The Barred Rocks of my youth were splendid birds with sparkling barred patterning and clear and vivid yellow feet and legs. I haven't seen their like in a great many years and I know many who are searching widely for them.

With some breeds such as the Barred Rock you will find producers maintaining separate flocks to produce male and female offspring. While these are now primarily exhibition breeders this is a practice with some long roots.

I haven't ducked controversy so far — so, why start now? This is not the way to build and sustain a working fowl breed. The exhibition hall always has and always will have an important role to play with purebred poultry, but the farm flock must hew to certain constraints to assure its survival in simple facilities and under management for optimal returns. Genotype and phenotype should always be reflective of each other and pursuing too exacting a goal for a single trait like color or pattern can adversely impact other, economically important traits.

On the subject of seedstock production the difference between "breeders," "propagators/multipliers," and "commercial" producers should be noted here.

At one time the livestock community was well and fairly illustrated with a pyramid shaped structure. At the very top, representing no more than three to five percent of the total number of producers who were the "breeder/definers." They were an elite of sorts that placed their emphasis on shaping and defining breeds and type trends. Below them in the structure was a bit larger group, the "propagators/multipliers" that took what those above had produced, propagated it in goodly numbers, and then offered it to the large base group of commercial producers.

These groups often blurred together a bit, and over time the concept of such structuring has faded a bit. I believe it still has great validity in that it should give us all pause as to what role we actually wish to undertake.

Due to the presence of hatcheries and show breeders, the structure within poultry production is a bit more complex, and roles conceded to hatcheries now were once held by some quite large purebred operations. Also present in the poultry community are a group of people termed "string men." These folks may be breeders of some varieties, keepers of others, and dealers in some birds. In an earlier day some string men would hit the show circuit with "strings" of several hundred birds, some homebred and some bought, and they would fill many classes at state fairs and other poultry shows.

They would buy and sell as they went. A most important role that they continue to play is as a source of seedstock and the ability to advise where certain birds might be found. They aren't exactly "chicken traders," but even those souls have a role to play. String men often remove surplus and older birds from the scene, may be a channel to the ethnic trade, and I have bought a fair number of really good birds from these people over the years.

Step one in making your start with rare and heritage breeds is to clearly define to yourself and to others (primarily your potential customers) just what you are intent upon doing with their help.

What has to happen now for these rare and heritage birds to survive in a meaningful way is to have a group of real, visionary poultry producers gather around them. Not merely handfuls of chicken keepers but serious producers who will care about genetic purity and type. But they must not produce solely for the dictates of the showroom. These producers must respect the birds' history, but be committed to building upon rather than trying to seek control over it. They must recognize that it is not rarity, novelty or even a purple-ribboned grandpa that gives a bird real, long-term value.

Breed Preservation & Flock Building

While certainly neither commandments nor dictates, the following are some thoughts to help frame a mindset that should be helpful in your efforts at breed preservation and flock building.

• At this point in time many of these rare and heritage breeds and varieties exist in quite small numbers and often, in quite an imperfect state. Thus, the first birds acquired may come with a fair set of flaws to be overcome.

• Some populations are in various states of restoration and may not be exactly one hundred percent pure in their breeding. This is not a sin but something that needs to be known and fully disclosed. A full outcross to a completely different breed followed by eight matings back to a pure one will result in pure status again in the offspring.

• Note that some populations today may not be where they were left fifty to seventy years ago, but may have even regressed further back from there.

• Some of the very closely bred groups will need special care to overcome a lack of vigor, low libido, low egg output, and other reduced performance factors.

• Try to pick breeds and varieties with your own skill level in mind. Even some fairly common varieties can be a breeding challenge. For example, the Rhode Island Red is readily conceded to be one of the most difficult breeds to breed true to type and with the correct depth of color.

More breeders are needed for Black Wyandottes and White Dorkings, too. As your skills develop then go looking for challenges.

Having lots of birds and lots of breeds won't make for a better breeder. Some of the very best work takes a lifetime with just one breed or a variety within a breed.

There are literally hundreds of chicken breeds and varieties that can be termed heirloom or heritage. There is no one "best" among them.

Faverolles or Orloffs can be that choice if you accept their complexities of type and color and potential performance limits. Few, if any, will look like the pictures in the catalogs, but remember — they are just the beginning point.

In the mid-Sixties I saw boxes of twenty-five baby chicks sent off at consignment austions to sell for whatever they brought, sometimes for as little as twenty-five cents a box. In the mid-Seventies that same quarter would buy an adult bird at our local sale barn. By the mid-Eighties many felt that the chicken had gone away forever, walled away inside high-volume confinement units.

About that time though, chickens were again taken up anew by the back-to-the-land movement of the moment. There were but a handful of breeds to be easily had from a small number of hatcheries and a great many of them didn't look at all like the catalog pictures when you got them home.

Two dollar a pound range broilers and three dollar a dozen brown eggs got a whole lot of people thinking about chickens again. Those prices were to be had largely near major cities and such demand has probably reached a plateau somewhat of late. At the peak of such demand I had an Old Order Amish friend growing out and selling as many as ten thousand broiler chicks each year. His sales grew through naught but word-of-mouth. He ultimately changed course because the business actually grew to challenge his Amish way of life.

Will we be selling one dollar, two dollar or even five dollar a dozen eggs ten years from now? I cannot say, but it would be wrong to get involved with these birds solely because of reports of relatively high dollar selling prices.

I am sure that if growth continues at any level near current trends we will see as many or more fifty-dollar breeding birds sold as we see right now. It will not be because of their scarcity, however, but because of their quality breeding and the improved performance that they can bring to other flocks.

A HATCHERY CATALOG PRIMER

CREATING THE PERFECT WISH LIST FOR YOUR FARM

It is the time of year when the hatchery catalogs are stacking up on the desktop. Although there are not nearly as many of them as there used to be, I can still wile away many an hour turning through their pages. I once wished — and believe there is still a need — for a primer of sorts, to be read right along with them . . . not because they are poorly written or deceptive, but simply because they try to showcase so much, often in the vocabulary of the established producer.

Pullet, poult, pigeon, partridge pattern, pouter, toe punch, pea comb, pearl, and *Plymouth Rock* (all eight-plus colors and patterns) are all part of that vocabulary. It can be quite confusing to the newcomer, and like any other wish book there are always temptations aplenty in a chick catalog. The first one to resist is that notion to order three of these, four of those, two of the red ones, and at least one each of those fuzzy-looking ones to see what they grow into.

We are starting our second decade with heritage poultry and have owned breeds from the Ameraucana to Wyandottes in four different colors. At one time our breed count stood at 23 — but at those numbers, *they* start owning *you,* and some pens don't really get the producer focus they really need.

From a box of 25 as-hatched baby chicks a truly great flock may grow. Still, of those 25, expect no more than 10 to be pullets, maybe one good "keeper" trio (one male and two females), and one or two more backup pairs. Although their elimination can be a tempting way to cut feed costs, experience soon shows the value of having one or two backup breeding males on hand at all times.

Job one should be to acquire the best-bred chicks possible within budget constraints, and to that end, do not hesitate to ask questions (*lots* of questions) before placing an order. You can't expect state fair winners in hatchery run chicks, not even if you're buying from show breeders. *Do* ask about the breeding behind the entire flock, the production history behind them, and any other questions you feel pertinent to the selection of birds with which to build your flock.

A great many hatcheries no longer maintain extensive breeding flocks, and with some of the rarer breeds, multiple hatcheries may be buying their eggs from the same flock.

We tend to be partial to some of the smaller hatcheries and those a bit off the beaten path. They are more likely to be hatching completely from their own flocks, and from long-held breeding lines, too. To find the locations of these hatcheries and a great many private breeders, your local Extension Office can help you acquire current copies of your state's Poultry Improvement Plan and National Poultry Improvement Plan directories. These are regularly updated listings of those testing for pullorum and paratyphoid in their birds.

HOW MANY?

Right now the very rarest and/or most in demand of the heritage breeds are usually available only in very small lots — sometimes just five or ten chicks per order. These small-lot purchases carry one very real caveat: they can be, quite easily, all of one sex. Even with the more popular breeds and where sexing is offered, the concept of buying three pullet-two cockerel or four pullet-one cockerel blends of breeds is seldom a good idea.

This is simply spreading things too thin. Almost assuredly, the one bird you will lose will be that lone cockerel — or he will develop with some real quality issues. With less than a score of chicks, you can also encounter real selection problems. If you must buy in small lots, seek to buy two or more such lots in season and even shop multiple sources if necessary.

Another aspect of uncertainty may grow out of those small lot purchases: "filler chicks." Most hatcheries and breeders will ship a minimum of no less than 25 baby chicks. This is to assure that the peeps remain warm and secure in transit. A few — a very few these days — may ship a minimum of 15 chicks in very warm weather.

If your needs or budget or the hatchery's ability to supply limits the number of chicks that can be sent at one time, the supplier will then add "filler chicks" up to their shipping minimum. They will be at a cost to you, a cost that can vary widely. You can pay the price and add chicks of breeds you choose, you may request a mix of their overruns (generally offered at a discount), or the supplier may fill out the box with whatever is at hand at a minimum additional cost.

With one order of Cuckoo Marans a few years ago, for example, we left the filler option to the breeder and were sent fair numbers of two other rare breeds. More recently, a group of chicks to replace birds lost in a shipping problem was rounded out with a slew of White Rock male chicks. At one time the larger hatcheries relied almost entirely on White Leghorn cockerels for filler chicks.

If you do decide to buy some of the rarer breeds, by all means order enough to create a viable breeding group. More and more often we see six to ten chick minimums for the really rare breeds. Likewise, if the budget can hold it, buy something other than discount filler chicks.

STUDY & RESEARCH

Peruse those catalogs and breeder lists carefully. Study them like a text, making copious notes as you go along. Once you have settled upon a breed, do some comparison shopping both for price and quality. If you haven't settled upon a single breed, buying and growing out groups of two or three breeds under consideration might prove useful. You may even have to work through two or three sources of supply before finding birds of the type and breeding that best fit your farm and your goals for your flock.

Over the years we have trialed several lines of Buff Orpingtons, for example, looking for the line that would work best here and for our chick customers. We have encountered problems with everything from small size and slow growth to poor lay and egg-size issues. In time we will find the genetic pieces to put together to create our desired flock.

The chick catalogs still fascinate me, just as that big Sears Christmas catalog with all of the toys once did so long ago. I even remember when Sears had baby chicks in some of their catalogs! Times have changed, but the fundamentals remain solid.

Originally published in Acres U.S.A.

CAPONS OR NOT?

One question I have been asked several times of late is, "What about capons?" I think this is because there is a growing interest in an alternative to the range broiler, or a second-line bird to be paired with it.

A capon is a castrate that is meant to be fed to a heavier market weight. It develops none of the aggressiveness or other secondary sexual characteristics of the intact male. I have even seen old photos of heavy capons mothering baby chicks.

Caponization is accomplished through a small incision in the bird's side, with a special tool used to extract the testes. It is not an overly difficult task, but is rather labor-intensive, as every bird must be caught and restrained.

The "roaster" may be a better alternative than the capon. These are older table birds, 14 to 18 weeks of age. If produced as males of the heavier breeds, they will yield a dressed bird in the five- to seven-pound weight range. They are no longer "fryers," but do fit the classic definition of a young and succulent roasting fowl. They can center those holiday meals for today's smaller families and those wanting an alternative to turkey or ham.

Males in this category will be fairly inexpensive as chicks. As intact males, they should grow rather quickly and should reach a handy harvest weight before their more aggressive nature can emerge.

These birds will, at least initially, be likely to sell in fairly small numbers, perhaps no more than one or two per customer.

This is a classic poultry product not often available today. Their best role may be as an add-on to a range broiler program, but demand should grow as word of their availability grows. They are a far more valuable bird than a spent roasting hen salvaged from the laying flock. One local producer markets a number of 12- to 14-week-old heavy cockerels for $7 each, and more during the end of the year's holiday season.

This is a new/old product from the family farm poultry producer and may need a bit of additional marketing at first. Orpingtons, Light Sussex, and other white-skinned birds could be marketed as "English" roasting fowl. Rocks, Wyandottes, New Hampshire Reds, Delawares and Dominiques were once America's classic Sunday dinner — and should be again!

The capon has a gourmet-fare history that can now be equated with today's faster-growing, better-fed young purebred male birds. The place to begin with them may be a box of 25 heavy-breed cockerels bought in the early fall. It is neither broiler nor turkey for roasting, but rather, something in between and available only from a relative handful of family farm producers.

Originally published in Acres U.S.A.

· CHAPTER 4 ·

THE HEN HOUSE & THE POULTRY YARD

SILVER SPANGLED
• HAMBURG MALE •

SILVER SPANGLED
• HAMBURG FEMALE •

SINGLE COMB BLACK
• MINORCA MALE •

SINGLE COMB BLACK
• MINORCA FEMALE •

I T IS SAID THAT CHICKENS WAKE UP IN A WHOLE NEW WORLD EACH AND EVERY MORNING. Little matters to them as long as they are kept dry, protected from drafts, kept out of the muck, and safe so that critters can't eat them.

This is all that is needed for satisfactory poultry housing. Would that it were as simple to do as it is to say. On my homeplace, the old horse barn was once home to a flock of half-wild bantams that literally lived off the fat of the land. They ate grain spilled by the hogs and cattle and supplemented this with insects and nearby greenery wisps of hay in the fall and winter. They nested up high in the old horse barn hayloft, roosted along its stall sides, and crowed the sun up every morning. They would hatch their young from roughly Easter to Christmas, and their solution to predation was to just do their best to outnumber 'em. Would that my layers and blooded stock now were that vigorous and self-reliant.

Over the years since my days on that old farm I have seen many chicken palaces and chicken gulags run by corporations. An image from my freshman year in high school stays with me still and tempers my thinking about poultry facilities to this very day.

I was a member of a Future Farmers of America (FFA) judging team for poultry then. At one district contest we were assigned a judging class of four, truly huge Buff Cochin roosters. We were told that they could not be handled, as they had just been sold for the then princely sum of fifty dollars per bird. That was "buying money for a used car" in those long ago days of my youth. What struck me most as we watched those truly golden birds was how they each calmly and contentedly drank from a used tomato can wired into the corners of their cages. An old tin can suited them just fine.

Let me set down something here that is an absolute iron tenet of any sort of livestock production. You never make a dime from the buildings on a farm, but rather with the birds and animals that pass through them. Money is made from the birds and this is where to invest money first.

That said, investment in good poultry genetics needs both protection and the support equipment to facilitate good production from them. Just keep in mind that it is not the equipment that makes the birds better.

To most people now, even many farmers, chicken equipment means environmentally controlled buildings, automatic feeders and waterers, mechanized manure handling, and even automated egg collection. This wasn't always so. Now even those with several hundred birds cannot afford such a big investment in housing and facilities.

Poultry housing was once the anchor of nearly every farmyard layout. Thus with heritage breeds today we should take our clues to their housing and care from history as well. An evolution in poultry housing and management practices has led the way to nearly all of the modern designs and systems now seen in all types of livestock production.

Down near our county seat is one of the last of the pre-pole barn-era farmsteads. The farm sets atop a finger of rolling ground and was laid out for laying hens and a small dairy herd. Those poultry buildings, painted in the classic green and white colors for such structures, are just as clean in their design and functionality now as they were in the days prior to World War II.

The main buildings are of shed-type design, fourteen to sixteen feet deep, roughly two to three times as long as they are deep, and are built facing south. The challenge for poultry housing of that era was to keep the birds laying in the cold season with its shorter daylight hours. To that end poultry sheds were nearly always faced south.

This turned them from prevailing winds and positioned them for maximum exposure to the thin sunlight of a Midwestern winter. Many had numerous windows glazed into the taller, south-facing side of the building. The wall might be filled with windows its entire length and from two to three feet above the floor to the roofline at seven to eight feet.

The goal was to capture as much of the heat of the sun as possible and to get all of that sunlight into the deepest recesses of the houses. Shutters might be closed or canvas curtains lowered at day's end due to the poor insulating qualities of glass.

DESIGNING A HEN HOUSE

The hen house design should be every bit as well thought out as the barn and the family home. Some design tips to consider include:

- Position it on a slight rise to improve airflow through and around the hen house during very hot weather.
- Our traditional Midwestern poultry houses have dirt floors that can foster some rodent problems. We keep on top of it with bait stations and believe that the dirt floors work best with a deep litter system.
- With deep litter, the floor is covered with four to six inches of bedding material such as fine or chopped straw. Fresh bedding is added as needed and wet spots should be forked up and out as needed. Late in the day a bit of grain can be sprinkled across the litter to encourage the hens to scratch the surface and keep it a bit fluffed. As the lower levels compress and begin to break down they will actually generate a bit of heat that may add to bird comfort during cold weather.
- Where cost is no concern, concrete or sand floors are recommended, but I question their practicability for the small flock. If you are rehabbing an existing building for chickens, go with what is already there.
- In very hot weather the roof and outside walls can be misted with water to reduce surface temperatures and tap into the evaporative cooling process.
- Each fall, interior walls and roosts can be whitewashed with a lime solution to improve light inside the building. This is also provides a natural checking of certain parasites.

I well remember forking out chicken houses early in the spring and late in the fall prior to changes in the seasons. We cleaned down to hard packed earth. The manure and litter went onto the garden or on the crop soil. It wasn't taxing work and was rather a pleasant way to stay busy on a cool day.

These older Midwestern poultry houses were the prototypes of the loose bird laying houses. They provided two square feet or a bit more of floor space for each member of the smaller breeds (Leghorns, Anconas, etc.) and up to four square feet of space for every one of the larger breed chickens. Inside, a twelve-inch by twelve-inch by twelve- to eighteen-inch nest box was provided for every five hens housed.

Roost space of twelve inches or so would also be provided for each bird housed. Removable dropping boards were often positioned beneath the roosts to facilitate house cleaning and sanitation. Roost sections were often hinged, so that they could be raised up and out of the way for cleaning or house depopulation. Four to six inches of trough space would be provided for each bird and one drinker or fount for each twenty-five birds.

Many of these older Midwestern poultry houses still exist in various states of repair along with other shed-type outbuildings that can be refitted for today's poultry flock needs. With a smaller flock oftentimes they don't even need a whole building dedicated to them.

I have seen many small flocks sheltered in a single, segmented section of a larger building used for multiple purposes. Birds should not be thrown in among other species and they should especially not be too close to hogs. Poultry should also not be roaming around in the twilight, roosting freely, and soiling farm equipment or supplies. There are few situations in a farmyard where at least a few birds can't be fit in if properly contained.

With heritage and rare breeds, most likely in the early years only modest numbers will be present. They are precious stuff and should be protected as such, but they were not bred to be tender, hothouse orchids. There are no heater houses needed here.

A great many heirloom producers now aspire to at least some form of partial range production. However, range production does not mean simply throwing open the chicken house door and letting them run until the foxes eat them or the semi's flatten them.

Chickens are really not very efficient users of browse and the reasoning behind putting them outdoors is not to capitalize on their strengths as a grazing animal. Out of doors they receive more sunlight, exercise more, have access to a greater array of feedstuffs from worms to weed seeds, have a more stimulating and satisfying environment, and are simply free to be chickens.

Ranging without any constraint or control may achieve some of the above, but it will expose the birds to greater predation and make it far more difficult to access them. On our farm are chickens, turkeys, wrens, pigeons, blue jays and cardinals. We value them all, but some need more care and involvement from us than the others.

Over the years some quite simple and other more elaborate methods and practices have been employed to safely bring chickens out of doors. On some farms, birds would be released into the farmyard late each morning following laying, gamecocks would be tied out with leg tethers, and birds literally allowed to live a "catch-as-catch-can" existence. Those that did survive were very hardy, very, very wary, and most often, wise and tough old roosters.

A common practice early in the last century was to position a poultry house at the center of four contained and seeded lots. The birds sheltered in the house would then be rotated through each of the four lots as the condition of the plantings growth dictated. This also helped to break the life cycle of some parasites.

An old neighbor of ours still employs a version of this with the several hundred broilers he produces each year. He has two simple wooden buildings of about eight feet by sixteen feet each. He has them on runners and can move them about easily with a moderate-sized tractor.

Each unit will brood four hundred or so chicks and our neighbor will produce at least two sets of birds per house per year. He will start his first set of chicks with the houses positioned in a fresh location each season. A location that chickens have not ranged upon for at least a year.

After about four weeks, with the chicks in good feather and no longer needing supplemental heat he will open one end door on the building. The young birds will then generally range out from that end of the building in a more or less fan-shaped pattern. They are let out each morning as a part of early chores and shut back in the house just before dark. Not the classic range situation, perhaps, but it is one that does get his birds out of doors and functioning most naturally.

Chicken tractors, portable grower pens, egg-mobiles, and even sun porches have all been about the compromise between getting chickens outdoors and keeping them apart from the things that live there and would eat them. One of the first and hardest lessons in poultry keeping is that people aren't the only ones who like a good chicken dinner. With the heritage breeds it can sometimes be a quite costly meal to serve to a predator.

There are some birds here that I would frankly hate to see risked in a totally free-ranging situation. They are either too rare, too valuable or too lacking in vigor for the challenge of free range. A few years ago, with grant assistance from the Missouri Department of Agriculture, we borrowed a concept from the British that has been useful in getting some of our rare breeds at least a bit more outside while holding them in a fairly controlled area.

With this funding we made a set of A-framed housing units that are thirty-three and a half inches in width and seven and a half feet in length. Half of each unit was enclosed on the bottom and the two sides with plywood. The other half was wrapped with two inch by four-inch wire mesh. Down the bottom half of each side went a twenty-four inch wide strip of octagon-meshed, baby chick wire.

These units will act as breeding pens for five to eight birds depending upon the variety. They will also hold up to forty growing chicks as they grow up to about four weeks of age. Beyond the chick stage, these units will grow out twelve to fifteen broilers or roasters. For us they provide that extra measure of security needed with very rare and hard to replace birds.

Our poultry operation is based upon the poultry yard concept, something else borrowed from those good poultry raisers — the British. The poultry yard concert is not simply a new take on your grandmother's chicken yard, but rather a systematic approach to anchoring poultry into an existing farmyard. The goal with it is to best utilize available facilities and geography to create a safe and efficient production site.

Basically it is a collection of small pens, runs, and freestanding units gathered around a central hub. These fit into our small farm's basic landscape and are tucked in among existing trees, adjacent to a water source, and facing away from prevailing winds. The small units fit well with our preservation-based breeding program that uses small breeding pens. Future plans are to consolidate two larger units to create a containment facility for a modest sized laying flock.

The hub of the unit, situated beneath a dusk-to-dawn light, is an eight-foot by ten-foot building (yes, a yard barn) that holds a cabinet incubator, early brooding units, workbench, and tool and supply storage. Beneath the shade of native trees around the yard are a couple of traditional, shed-type poultry houses, a number of A-frame poultry arks, several freestanding units that are basically rehabbed rabbit hutches and dog kennels, and a few brooder/grower units with attached sun porches.

Much of it is cobbled together, fusions of many parts, and is even rebuilt from time to time to meet a new need. Our goal is not beauty, but to keep breeding birds and small lots of growing birds solidly in place, easily accessible, and safely protected from anything that goes bump in the night and likes to eat chicken. There are tops on everything — as a free roaming, free booting rooster can set back breeding plans for several weeks.

There is a lot of flex inherent in this concept and it can be brought to a great many farms and smallholdings. The many smaller pens make it easier to set up and regroup matings throughout the breeding season and to evaluate the young birds as they grow. Their elevated nature is also helpful with vermin control, waste handling, and predator prevention. They are also just a whole lot easier for us old, stiff people to work around when they are raised off the ground.

Much of our equipment was bought second hand and started life as containment for all sorts of other critters. Many are largely just boxes on legs to which we have attached all-wire rabbit hutches with fender washers and screws. These "sun porches" are one piece and can be had or built in many sizes. The washers and screws make them easy to disassemble and adjust. A few we have rebuilt from the get-go and they just don't fit into any recognizable category.

One of our brooder/grower units actually began life as a shipping crate for a coffin. Its array of side handles still draw a great many second glances from unsuspecting visitors. With my sore knees and older than middle-aged back, none of which bend easily, most of our cages are elevated to at least waist height (and even a bit higher). It is a step that also keeps the birds more accessible and easier to visually inspect.

The sun porch concept is incorporated into a great many of our breeding and growing units. It is an old-line poultry-housing concept that still retains many sound management principles. These raised runs with heavy mesh bottoms make it possible to get growing birds and breeders outdoors in all kinds of weather. Wastes fall through to the ground below to be retrieved for the garden or compost heap and the birds are less likely to pick up parasites. It is also an outdoor containment method with a measure of added bio-security in this time of concern over possible Avian Influenza problems.

This was a practice initially seen with growing turkeys. The birds so contained have an enclosed area for sleeping and laying and can take a turn out on the wire run even on rainy or snowy days.

Many books have been written strictly about one or another type of housing for birds. Realize that what works for us may not work for someone else. One of our friends has a laying house made from straw bales, several others have egg-mobiles (chicken houses built on wagon frames), while others have hog houses that have been reworked for poultry housing, and all sorts of pre-fabricated units have even appeared of late.

Regardless of building styles, most small producers will be working with what is called "cold housing." The building may be heavily insulated and have tightly sealed sides and floors to prevent drafts, but will have no supplemental heat source. The birds will gradually acclimate to the changing seasons and as long as they are protected from drafts, dampness, and direct sun they should fare well in these simple quarters.

A problem to be mindful of in very cold weather is freezing damage to combs and wattles. It is generally not fatal, but can be disfiguring. Especially vulnerable to such damage are the single combed birds with very large comb and wattles. Comb points and portions of the wattles may even die and slough away. Until fully healed, the birds may run a low-grade fever and this increase in body temperature can render males at least temporarily sterile.

A few years ago a hard freeze in early March caught us with breeding pens already set up and many males set out in simpler housing. Comb damage was extensive around the poultry yard and some pens did not produce fertile eggs until late June of that year. Males are far more vulnerable to this sort of thing as they roost with their heads up — unlike females that will tuck their head under a wing on a very cold night. In many of our small units we do not provide roosts and believe that the birds may actually fare better by hunkering together in one corner during cold weather.

This event somewhat hastened our move toward more rosecomb breeds. They do seem to fare better in the Show-Me state's always-unpredictable weather. As I write this we have just finished one of the warmest Decembers in state history, but morning temperatures right now are in the single digits. However, there are still Leghorns and other single combed birds here. Now there are measures that we can take to protect their combs and wattles.

On nights when temperatures are expected to plummet, collect up the males and heavy combed females and liberally cover their combs and rattles with petroleum jelly, menthol rub or similar heavy salve. Be careful not to cover their nostrils or get this in their eyes. This treatment may have to be frequently reapplied. Some poultry farmers will catch each breeding male on a cold winter evening and place it in a fifteen dozen egg case or similar, large, easily-closed, cardboard box. These boxes will help the bird to better contain its body heat, can be placed adjacent to the roost area or even removed to a cellar or other area where temperatures won't fall too sharply.

Always remember that birds experience their housing environment at different levels than you. Humans experience the environment based on his or her shoulder level. The birds live at floor and roost level. This is where you need to get down to look for draft sources, predator routes, and other potential housing problems. In mild weather during the breeding season we will sometimes use the larger, all-wire rabbit cages to hold pairs or trios used in side matings. These can be suspended in a building with a higher ceiling, in old hutch frames or even between steel posts positioned beneath large trees and covered with simple plastic tarps. The thirty-by-thirty by eighteen-inch (or larger) cages are really good for containing birds being used in pedigreed matings. Be sure that headroom is adequate for the males to stand and breed. We keep birds in them for a relatively short time and cover one-third to one-half of the floor with thin plywood or other solid material so that the birds are not always standing on the wire mesh.

Good housing and equipment are meant to facilitate the keeping of chickens. The facilities should never be a big profit eater nor an artificial crutch or prop for birds that would not live and thrive otherwise. Dollars spent on housing and equipment are to optimize performance and production from the birds.

Whether containing one bird or one thousand birds, the key is to keep them comfortable and as free of stress as possible. They need fresh air and sunshine and to be in an environment that contains them safely and cleanly. Their housing should keep them held in a way that you can fully observe their condition too. This last point cannot be made often enough.

For example, every few years here during the late spring/early summer brooding season there is a period of damp, cool weather. During this time a set of chicks face the prospect of outgrowing their primary brooder space.

When they are too crowded their growth will slow, then they will begin to lose bloom, and you might even start to lose a couple of chicks a day (generally the rarest and most costly ones too). Old-timers may call this "brooder ill" or something similar. Ultimately the solution is to swallow hard and move them out into a more open and larger grower pen. Perhaps a few more will be lost, but just chalk it up to "natural selection" and then watch the sturdy ones gain ground and begin to bloom again.

Literally thousands of dollars can be spent on equipment and poultry facilities, but once the plans move beyond the basic tasks of keeping them safe and comfortable it is largely then all for ego and just plain window dressing. Good facilities generally reflect a good producer — but the producer should also realize that big money investments should always be in the birds and the feedstuffs to fuel them.

According to Morrison's classic text on livestock feeding, *Feeds and Feeding,* an acre of good pasture will carry two hundred hens or three hundred growing birds that are also receiving a well-formulated ration. During the course of a year a single adult bird will consume fourteen to twenty-one gallons of water. Water fountains, troughs and cans vary greatly in design and capacity, but what is the most crucial is that they do not allow the birds to roost upon them and defecate into the water. An old rule of thumb holds that each adult bird will need three to six inches of feed trough space and each young bird will need one to three inches of feed trough space as they grow and mature.

Range production is essentially the containment of adult and well-feathered young birds in a manner that enables them to exercise and forage for a small amount of their daily diet. In most parts of the country it is also a practice that is quite seasonal in its nature.

A young bird ready for life on the range is generally four to five weeks of age (by this time many confinement Cornish crosses are ready for harvest), is well feathered, and has been hardened off from needing any form of supplemental heat. Most are moved to portable grower units or chicken tractors that will hold them fully contained while providing about two square feet of pen space for each bird. Such units are normally moved once or twice each day to keep the birds on "clean" ground and to better distribute their wastes.

The typical "chicken tractor" range unit is ten feet by ten feet and is made of lightweight materials for easy portability. Usually it has one-third to one-half of the structure enclosed for night and rain shelter. Special attention is required to prevent wind damage and to provide protection from predator attacks. Although it will add to construction costs, it is far better to enclose these units first with a narrow-stayed wire mesh. One inch by two inch or even one inch by one inch, heavy gauge wire mesh will help to foil predators and make such units more durable than those enclosed with octagonal chicken wire.

Canid predators will go to the corner of a pen to attempt to breach it. To counter this, some are beginning to experiment with round pens and structures for broilers on range. A top cover is also a must to keep the birds safely in place and protected from winged predators.

Layers and breeders on range are sometimes allowed to range somewhat more freely and their primary protection and containment lies in the area perimeter fences. The ideal fence would be horse high, bullhorn strong, and baby chick tight. Sadly, it would cost many dollars a foot to buy and erect.

The long time standard for a poultry perimeter fence is to have a height of five to six feet with the top eighteen to twenty-four inches flared outward. It would have two by four inch or smaller mesh with the smaller mesh definitely along the bottom two feet. Also an additional two-foot wire apron should be extended out from the fence at or just below the ground level. This fence can then be reinforced with one or two strands of smooth electric wire at four and eight inches above ground level. The fence now contains measures intended to thwart both digging and climbing predators.

A fully charged electric mesh fencing material of forty-four to forty-eight inches in height is now available in one hundred sixty-five foot lengths. Simple, step-in plastic posts are built into its entire length for quick and easy erection. It can be bought from fifty cents to one dollar (or more) per foot and should do well containing most of the heavier, less prone to flying breeds. This fence will require a fairly powerful charger unit to power it. Also, closely mow the perimeter before erection to prevent plant growth from shorting out the fencing when in operation. If used at any distance from the farmstead, the charger will have to be either a battery or solar powered model. If the electrical charge can be kept strong it will turn away some of the most aggressive predators. Two, three or four of these live wire strands can be linked together to form large and irregularly shaped enclosures. The greatest strength of this type of enclosure may well be the speed and ease with which they can be taken down, moved and re-erected.

Most of the farmsteads of my youth had both a "chicken house" for the layers and adjacent to it might be a small brooder house with the capacity to brood a couple of hundred chicks. Adjoining many of those layer houses would be a drylot that would allow the birds out into a fenced yard nearly every day. It was not a true range environment, but if managed well, it did get the birds outside to receive more exercise and fresh air. This setup could help simplify things for the producer during the harsher seasons.

Such a lot would be enclosed with a tall fence such as described above and provide two to four square feet of ground space for each bird contained in the chicken house. This is probably best used with flocks of one hundred fifty birds or less. Begin by leveling the lot area. Over it place a two-inch layer of stone, then a layer of two to three inches of cull lime and then a one to two inch layer of sand on top. As needed, the top couple of inches can be skimmed off as a sanitary measure and replaced.

Here in Missouri where there are a lot of clayey soils, many of these lots are simply compacted earth that has built up over the years. The tops can be skimmed from time-to-time as a sanitation method as long as the grade remains above the soil line. A tight sod maintained around the pen will provide a natural filter for any rain runoff coming from the surrounding area. Such a lot can be dusted with straw in extremely wet weather and books of straw can be used to fill low points.

Green and white are the classic colors for painting poultry housing units in Missouri. Into the housing should go nests, roosts, simple light fixtures, feeding and watering equipment. Light is provided to encourage winter laying and a simple porcelain light fixture may be adequate for this in a house with a small flock.

For buildings that are ten by ten to ten by sixteen feet in size, a single, forty- to sixty-watt bulb in the center may be all that is needed for adequate flock lighting. The key to layer or breeder house lighting is to maintain the same number of sunshine hours as Mother Nature would provide on the longest day of the year in your area — the first day of summer. Set the light or lights on a timer so they come on in the early morning hours and go off at sunset. The lights should not be turned off in the house after sundown with birds still off of the roost.

Perhaps the most basic element within the hen house is the nest. Here the egg that launches everything is laid and all of the natural processes begin.

I have seen nests that range from bare-bones simple to throne room elaborate. On the simple side, some are little more than packing crates turned on their side. At one time eggs were widely transported in two section, thirty dozen capacity boxes made from lightweight wooden sheathing. Numbers of these egg crates would be stacked together in the henhouse and after a year or two of use, they can be removed, burned and ultimately replaced with fresh "nests."

Wall-mounted nests each twelve inches by twelve inches by sixteen inches (or a bit larger) are often homemade using one-inch lumber or bought already made — manufactured from sheet metal. These are generally made in units of four, six, eight or ten nests each and can be taken down for cleaning or repair. Some are now even made of plastic. They have a beveled top to prevent the birds from sitting or roosting atop them.

Nest boxes sometimes seem to have been little more than an attempt to make the best of a bad situation. We have had hens that would steal away to the farthest corner to lay, some that would lay seemingly wherever the mood hit them, and other hens that could pile three deep just to lay in their favorite nest. Open-fronted nests can also contribute to problems with egg eating. This type are not always kept clean, cannot be used for broody hens, and will provide no verification as to which hen laid which egg.

To keep the eggs cleaner and reduce problems with breakage, nest bottoms can be padded with a variety of different materials. Chopped or long fiber straw, ground corncobs, hay, and other farm-produced fibrous materials have been used for this purpose. They work well as long as they remain in the nests, but all too often the hens quickly scratch them from the nests. You can now buy rubberized nest pads that lay in place. Many people make their own nest pads from recycled indoor or outdoor carpet squares. These can be removed, cleaned and treated with a disinfecting bleach solution or simply discarded when too worn or stained.

With the very dark pigmented eggs, the shell color is not fully set on the shell until shortly after the egg has been laid. In fact it can be smeared if handled too quickly after laying or by contact with certain nesting materials or surfaces. Pads and stiff-fibered squares, if kept clean and pliable, are probably the best choices for use with these breeds.

Egg eating can be an especially costly problem, not just because of eggs lost, but also because of other eggs being badly stained with yolk and white from the broken eggs. Such eggs are hard to clean and very poor choices for incubation.

There are near countless folklore cures for the problem of egg eating including my grandmother's favorite of loading a blown or cracked egg with red pepper and putting it back in the nest. The best method is perhaps to simply identify and remove the egg eaters. Look for birds with beaks, faces and breasts stained with egg yolk. Some will give them something else to peck at to keep them occupied such as a head of cabbage suspended at or just above their heads.

With the rare breeds absolutely every egg counts and we have found that a scaled down version of what has come to be called the "colony nest" works best for our small breeding pens. We have a set of four foot by four foot by thirty-inch plywood boxes with hinged tops that we place in houses with our larger breeding and laying groups. One of these units can easily serve up to twenty or twenty-five hens and they can be made in even larger dimensions for larger numbers of birds.

Admission into and egress from this box nest is through a single eight by twelve inch opening in one front corner. This unit tends to discourage egg eating, as the birds cannot see into the areas where eggs accumulate. The birds tend to enter, retreat to one of the back corners to lay and then exit quickly through the opening. The one source of light that they can see when inside is at the entrance. We cover the bottoms of these units with a layer of straw, wood shavings or straw over shavings. Cedar shavings tend to deter some parasites if it is replaced often and is kept fresh and clean.

Nests should be cleaned regularly and often. Before replacing the fresh nesting material or cleaned pad, dust the nest bottom and sides with a bit of agricultural lime or even an appropriate dust if mites have been a problem. If possible, set nests and roost poles outside once or twice a year to catch at least a day or two of the naturally disinfecting sun's rays.

The wall-mounted units generally fail due to abuse and neglect. The two most common points of failure are usually the step-in boards and nest bottoms. The bottoms may corrode or rust away and can be replaced with properly sized squares cut from sheet metal, plywood or hard board. The step-in boards are generally pine slats and are fairly simple to replace.

Before leaving the subject of the hen house and its attached fixtures there should be a bit more detail on how best to thwart the many and varied predators that can assail the birds contained there.

THWARTING HEN HOUSE PREDATORS

We have seen or tried many measures to thwart poultry predators; here is a list of those that have proved to be at least somewhat successful. Just don't rely on only one to keep your birds safe.

A Great Pyrenees or other type of guard dog that has been imprinted on your flock and farmstead, just can't be beat for keeping your flock safe. We have a largish Rat Terrier that patrols our poultry yard now.

Ring the poultry yard or range with multiple strands of smooth wire, electric fencing, set at heights of two, four, six and/or eight inches. Two charged strands of electric fencing will turn away a good many of the smaller, closer to the ground predators. The wire and fence needs to be inspected often to see that all is working well and the lines have not shorted out from contact with grass or fallen limbs.

Long, very narrow, open-topped runs seem to do a better job of deterring winged predators than wide runs. Pen tops will eliminate many predator problems. An in-between measure is to zigzag strands of smooth wire across the top of the pen. If they flex a bit and vibrate in the wind — so much the better.

Thin, long strips of silver mylar cut from old balloons, tied out and left free to flicker in the wind will distract and deter some hawks.

THE SIZZLE CHICKEN EXPERIENCE

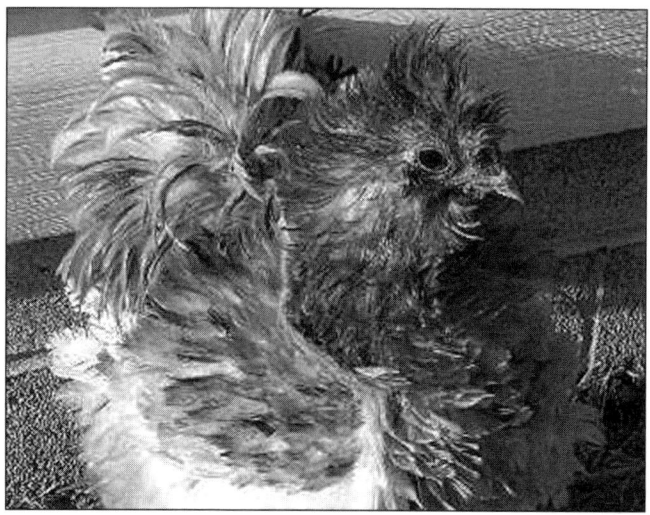

Like many other people who thought growing some of my own food would be a great endeavor, I bought a few chickens at the feed store and my education began. I quickly discovered there are more breeds than Rhode Island Reds and Barred Rocks. I happened onto an Internet forum where I could get advice on how to properly feed and house these little chicks. I discovered the vast array of chickens were from all over the world and some with names that were difficult to pronounce. There were even birds that had crazy top knots and feathers all over their legs and feet. Almost immediately upon seeing my first picture of a Silkie, I knew I had to have some. I also fell in love with Cochins. Then I discovered the curly feathered Cochins, called Frizzles.

There was a gal on the forum that had some Silkie/Frizzled Cochins cross birds for sale so I figured I would kill two birds with one stone (who perpetuates this saying anyway?) and she sent me three young birds. This was the beginning of my journey in breeding those sweet birds that have since come to be known as Sizzles.

The three little ones started to grow and before long I knew I had a little cockerel. The others, I concluded, did not have a big red comb and were smaller so they must be pullets — if it could always be so easy to tell the

difference. Early on it was clear that I had two frizzled birds and one smooth. I was not at all sure why that was, so the search began for information on breeding Frizzles. Again the Internet provided a wealth of information.

I learned that the mating of two frizzled birds results in some of the offspring having a double dose of the frizzle gene or what the genetics primer calls "homozygous." The birds that end up with the double dose will have deficiencies in their feather structure that causes breakage and in some cases just plain sparse feathering. While some of the birds from this type of breeding will be excellent frizzles, twenty-five percent will have the defective feathering and will have to be put down or cared for in a special environment. Not being that interested in destroying twenty-five percent of the chickens I hatch, I decided the best way to approach breeding was to breed a smooth bird to a frizzled bird. The results of this type of breeding will produce fifty percent smooth and fifty percent frizzled birds. Oh, I said to myself, that is why Bad Hair Day, my little cockerel and Buttercup, had frizzled feathers but my little Runt was smooth. At least now I could go forward with a plan to breed my own birds.

As time went by, Bad Hair Day learned to crow and soon became a little tyrant. He was not much fun to have around at all but I bred him to my little smooth feathered girl and the first offspring soon came along. I knew at this point that Mr. Bad Hair Day would need to find a new home. He attacked anything and everything that moved in our yard. I then began my search for a new rooster. A friend had an extra Silkie rooster in the color I needed so off to Kentucky I went for my new rooster. The only thing was, I now had a different breed all together. How would that affect the breeding program I had just begun and what is now called the Sizzles?

Back to the Internet I went, to learn what I could about Silkies and what affect if any, would the frizzle gene have on the already unique feathers of a Silkie. I could find very little information on Silkies and frizzling. What I did find was a whole new gamete to my already limited knowledge of genetics. Silkies are considered to have 9 unique characteristics that distinguish them for other chickens. Primarily the feathers do not have barbicels that cause the feather to hold together along the shaft. They have blue ears, five toes, feathered legs and feet, crests, a walnut comb that differs from what many recognize as that on a typical chicken. They have black skin and bones and some have beards.

Oh well, I certainly could not worry about all those genes bouncing around inside one little chicken, so I proceeded to breed my Silkie to the frizzled and smooth girls. Now the offspring that were produced were about fifty percent Silkie feathered and fifty percent with barbed feathers. To complicate the matter even more, half of the Silkie feathered birds were frizzled as well and half of them looked like a regular Silkie. Wait a minute here. I wanted frizzled birds with barbed feathers. This is where understanding genetics helps. Without going into detail about Mendelian Laws, I will simply say that when you breed two types of birds together, dominant genes and recessive genes will dictate the appearance of the offspring. In my case, with all the various unique traits that Silkies have, obvious basic differences become apparent right up front. I am on my fifth generation after the first cross and with each generation I find that more variations in comb, leg feathering, ear color; etc., pop up.

I have used a pure Silkie rooster on frizzled barbed birds for each generation to this point. This breeding season I will finally breed Sizzle to Sizzle, a frizzled bird with barbed feathers, to a smooth bird with barbed feathers. Since many of the different characteristics of the Cochin and the Silkie are recessive genes, I expect to see more birds with a Cochin type comb, four toes, light skin and rounder bodies. I have found that when I breed a bird with five toes to one with four toes, the offspring will have five toes. Once I breed those birds to other birds with a similar genetic background then both sides have the recessive gene for four toes and I will end up with some that have four toes, some with four toes on one foot and five on the other. The same principal holds true for any recessive gene present in the birds.

The work continues in developing a new breed of chicken. How and why I am involved in the process still evades me. I just love those curly, crested, feather legged bundles of fluff. I have learned so much and it has been my pleasure to share my experience with anyone who is interested. Possibly it would have been easier to take a class in genetics and do the charts to forecast the outcome of my breeding program but it would not have been nearly as much fun. The good news is that with continued breeding and improving on each generation eventually we will see a Sizzle bird being shown at chicken shows and ultimately accepted by the American Bantam Association as a recognized breed.

Jaynese Knight of Knight Chicks who can be reached at jimnjay@farmerstel.com *or visit her website at* www.knightschicks.com. *Photos taken by Jaynese Knight.*

Supposedly an owl cannot fly through a strobe light. I have seen these for sale for a very few dollars in some garage sales.

Pen sides that flare out measurably at the top may discourage some climbing predators.

All birds on wire mesh runs need a solid sided and solid floored adjoining structure into which to retreat when they are threatened. It should be deep enough that raccoons cannot reach into it.

Keep the poultry yards clean, free of spilled feed, and don't pile lumber or other items on the ground near the birds.

Regularly maintain bait stations to control vermin.

Live traps and certain kinds of baits can be used to deter larger predators. Use something sweet or fishy to draw wily raccoons into the live traps.

A BETTER FEEDER & WATERER

For those seeking the fast track to fame and fortune, may I suggest forgetting about designing a better mousetrap. If you sincerely wish the world to beat a path to your doorstep, then what is needed is a truly waste-proof chicken feeder. Nothing else has so occupied the minds of poultry owners, in my lifetime. Chickens flip and scratch away a small fortune in chicken feed each year and little has been found to overcome this flaw in the nature of chicken.

A feeder or waterer design is deemed a success if it keeps its contents fresh, clean, and readily accessible to the birds. With our small groups, the most successful choices have been designs that are decades old and basically work by filling from the top and allowing the birds to eat or drink from the bottom.

For us the most efficient and feed-saving feeder choices to date have been the hanging feeders that require the birds to reach up and in, all the while working a bit for their dinner. We even use small hanging feeders for chicks as young as just a few days old. A second feeder modification that seems to save feed is one that requires the bird to reach its head through a rather small opening and then down to the feed. Beak flips and headshakes are discouraged and when they do occur, the feed particles are better contained.

Today there are plans aplenty for making your own hanging feeders from salvaged five-gallon plastic buckets with tops. With our breeding flocks and older growing birds we use manufactured units that are essentially galvanized metal tubes to which a shallow feeding pan has been attached. Until quite recently such units were available secondhand for just pennies on the dollar, but now the older ones with feed company logos on them have become collectibles.

We have had fair success with another, old-line concept — the "reach through" feeder. These are simple, shallow troughs that clip to the outside of a pen or cage. The birds reach their heads through small openings in the sides to eat from the troughs. With this and all other designs be wary of feeders with nooks and crannies where feed can build up and spoil. Such spoilage can cause health and sanitation problems as well as waste.

Some of the best small pen feeders we have had were old baby pig feeders given a second life in the poultry yard. They were made of galvanized metal or plastic and were quite solid and able to stand up to the young pigs and thus have served us well. They are designed to fit against walls or into corners.

Inventive producers have taken short lengths of four to eight inch diameter PVC pipe and created some most interesting poultry feeding and watering equipment. When capped they can be used to create corner fitting or side mounting feeder units. Split into half lengthwise they can be made into many, distinctive trough-type units.

Still around are a lot of those ancient poultry feed troughs that are elevated on metal or wooden legs. They have flip-up, stiff wire covers that are supposed to keep the birds out of the trough and prevent feed loss from scratch-out. They are quite well made as attested to by their enduring nature, but the rails for the birds to stand upon are often broken and the legs damaged from being buried in mud and manure.

A lot of baby chicks are started with feed sprinkled on paper towels right on the brooder floor. Others will expose them first to feed sprinkled in shallow, cardboard box tops or shallow plastic trays that are actually patterned after such box tops. Within a couple of days they can then be given feed in the low, covered troughs or round feeders screwed to fruit jars. I like the latter.

Stacked atop our grower and breeding units are all manner of feeders. They range from old pig feeders to hooded troughs to homemade units that defy simple description. None are perfect for every circumstance, but they cost very little. Although many are not as good as I wish they might be. Most are small and to encourage good feed consumption, most have to be topped up often to keep the feed fresh and appealing.

Waterer options for the small flock range from the gravity-fed bell waterers, double wall units, simple drinker cups and the aforementioned tomato can. With the simpler cold housing, perhaps the greatest challenge is to keep adequate water before the birds in very cold weather. In our system the best options for this have been cage cups with inward tapering sides and simple, black rubber tubs of various sizes.

In cold weather I will water at least twice a day and may even use warmed water. Most well water actually comes directly from the tap at 55 F. and doesn't need heating. I will water in the morning and then again in the late afternoon — trying to keep the birds on a regular schedule.

Late afternoon care in cold weather helps to get them ready for the long, cold night. A good drink of water and an offering of fresh feed will fuel them for the night ahead. Before each new filling, the tapered cage cups can be slid loose from the cage and a sharp rap to a solid surface will cause the ice to pop free from the cup. With the black tubs it is a similar matter to flip them over and stomp a couple of times on the tub's center. I often also carry along a small rubber mallet to tap on the sides and bottom of any ice-encrusted waterer.

In warm weather, feeders and waterers can develop sanitary problems. The smaller units may be dipped completely into a disinfecting solution of water and chlorine bleach (one ounce of bleach per one gallon of water) once or twice a week. Larger units can be swept clean with the same solution once a week or so. Adding a cap of hydrogen peroxide and/or one to two ounces of red cider vinegar per gallon of water a couple of times a week to the drinking water will freshen the units, improve water consumption, and boost the birds.

Abundant, fresh and clean drinking water is a most crucial feedstuff. Without it digestion will breakdown and other bodily systems will falter and then fail. Watering equipment, whether simple or complex, needs to keep the water clean and fresh to foster good levels of consumption. In warm weather we may replace the water two or three times a day to keep it fresh and appealing. For poultry there are a number of self-watering vessels that hold from one to ten gallons of water. They keep the contents clean and fresh while providing easy access for the birds. In these the water flows into shallow lips or recessed fountains.

These units are usually made of plastic or galvanized metal and either fill through a top opening or by sliding off an outer covering on the double walled models. The metal units can be placed atop electric heating units for use in very cold weather. There is a school of thought that holds that warmed water may actually improve bird performance in cold weather.

Many years ago we owned a three-gallon waterer the likes of which I haven't seen since. It was encased in what we used to call "cream can tin." It was double walled and very heavily insulated. We filled it by laying it on its side and pouring water into the reservoir through the recessed fountain in the side. The insulation gave it great weight and the water would flow freely from it in all but the very coldest of weather.

BROODERS

Whether you hatch your own chicks or buy them from off-farm sources, the purposes of the brooding process remain the same — to keep them safe, warm and dry. The brooder is the real starting point for their young lives.

A baby chick doesn't have a lot of body mass and can do very little to generate anything in the way of true body heat. In the first week of life they will need a draft-free environment with an air temperature of around 95 F. That temperature can be lowered by five to ten degrees each week as surrounding air temperatures allow and until the chicks can handle warm season outdoor temperatures — when they are about five weeks of age and are in good feather.

Some of the darker feathered heritage breeds such as our Rhode Island Reds may actually be a bit slower to feather. Late and early-hatched birds will need a longer period of heat supplementation, also. You can gradually harden them off of supplemental heat by turning off the heat source during the warmer times of each day and steadily extending their time spent without added heat.

Perhaps nowhere else will you see the creativity of the small farmer better employed than in the creation and development of brooder units for baby chicks. I have seen small lot brooders made from things as simple as old aquariums or little, plastic enclosed cat litter boxes. On the other extreme, there are stackable, box brooders that will each hold up to one hundred chicks for up to two weeks. They can cost over two hundred dollars per unit and will have to be set up in rooms or buildings where the air temperature stays above 55 F.

Actually the more stable the surrounding temperature and environment the better most brooder units will perform. Ours are in an outbuilding or are freestanding units and thus we don't start chicks earlier than mid- to late-March. Early brooder houses measured between six by six feet to ten by ten feet and were heated throughout with kerosene or other stove-type methods.

These were replaced by hood-type brooders that created a comfort zone for the chicks beneath the hoods. The hood brooders initially burned kerosene and then later were run on electric or propane. These are still used to start chicks in the large, colony houses that also have other sources of supplemental heat. On most farms where chicks are started in small lots and space is limited, this is reflected in brooder design and setup. Our primary starting brooder began life as a two section mating pen and nesting box for large breed pigeons. It is a two-segment box of eighteen inches by eighteen inches by forty-eight inches that I turned on end. The top was enclosed with one inch by one inch wire mesh. Each half will hold one of our weekly hatches for a few days until they are strong enough to be moved to a larger brooder/grower that is generally a freestanding unit.

The box costs a very few dollars. It is trouble-free to scrape and disinfect, simple to move and easy to expose to the naturally cleansing rays of the sun. The solid bottom safely contains the chicks, will not cause spraddled legs to develop, and is draft-free. Likewise, the solid sides surround and extend well above chick level. Its heat source is a heat lamp suspended some eighteen to twenty-four inches above the chicks.

Commonly seen on many small farms now is a reworking of materials at hand to create a brooder unit adequate to house chick lots of fifty to one hundred head until they are three to five weeks of age. There is often an enclosed, heated zone at one end and a more open, meshed floor part of some sort in the other. Often seen in these recycled units is an oblong water tank that is past its prime, but will hold chicks nicely in a secure environment. These may no longer be water tight, but offer a solid bottom, draft-free sides, easy access, good capacity, and can be heated with a single suspended heat lamp. They can be capped with wire mesh or old window screens and will hold a good number of chicks in a clean environment that is easily tended.

A simple, home fashioned version of the commercial box brooder described above can be made from one of the larger plastic storage boxes with a snap-on lid. Such a box in a handy twenty-four to thirty inch deep size can be bought for a very few dollars, is easy to clean and disinfect, and will last many years with simple care. Depending on the size they can hold twenty to thirty standard bred chicks up to a couple of weeks of age, and can even be stacked into simple frames made for them. The only major modification to be made to them is to cut a large square or rectangular opening into the top and cover it with fine wire mesh. The heat lamp will shine through this mesh to the chicks below.

These are heavily used in our area. A bit of litter topped with paper toweling is used to prevent any leg straddling. These can be positioned on shelves or racks in outbuildings, basements, and the like. The heat lamp will be more than adequate anywhere air temperatures around the boxes don't go below fifty or fifty-five degrees F.

I have even seen a great many of these boxes set up in back bedrooms or second bathrooms with quite good results. Litter is changed daily and the birds are moved out as soon as weather allows. A friend has a frame in his sun porch that holds six of these boxes stacked side-by-side and three high. In fact, it is a fairly common joke in our area poultry circles that it is well past time to take the young birds out when they start roosting on your headboard or shower rod.

A few folks still do maintain a small brooder house or brood under a hooded unit. These must be closely monitored when in use. Chicks are held in them with round barricades made of cardboard or lightweight metal. If brooding in a small building, round off the corners with flashing or other metal to prevent the chicks from piling up and smothering in the corners. Chicks will instinctively move away from the heat source, especially if daytime temperatures begin to warm. Old hands would sometimes paint the brooder skirts black or dark red to kill the glint that might cause the young chicks some stress.

Most brooder units on smallholdings are warmed with suspended heat lamps shining down on the backs of the baby chicks. Raising or lowering the lamps or changing the wattage of the bulbs in the reflectors regulates the temperature. There should always be a minimum of eighteen to twenty-four inches above the chicks and the litter on the brooder floor.

Most everyone is familiar with the 250-watt type heat bulbs, but we have found that the 125-watt types work every bit as well in all but the most extremely cold weather and they don't reach as deep into our pockets when in use. Some also favor the red-tinted heat bulbs, as there is strong evidence that these will reduce the incidences of feather picking and cannibalism in brooded chicks.

As the chicks grow and/or the weather becomes milder, you can even use lower wattage bulbs. In very warm weather we have even gone down to forty or even twenty-watt bulbs simply to provide the limited light needed to encourage the birds to eat and drink during the night. Do not use the soft white light bulbs as they generate relatively little warmth. We normally keep on hand an assortment of clear glass bulbs from forty to two hundred watts throughout the brooding season. Upon the recommendation of a friend we have begun experimenting with the bulbs sold for use in reptile cages with mixed results so far. They provide some heat and a light wavelength that is supposed to be comparable to sunlight.

USING HEAT LAMPS DURING BROODING SEASON

A certain amount of caution is always to be advised when heat lamps are in use.

• Never plug more than seven of the 250-watt bulbs into a single, dedicated circuit intended for their use.

• Do not suspend the lamps by their cords or with any lightweight or flammable material such as baling twine. The best choices for suspending heat lamps are light-gauge chain or smooth wire.

• In taller units the lamps can actually be positioned resting atop the all-wire sections of the brooder cover. Do not allow lamps any contact with wooden or plastic surfaces.

• The reflectors and bulbs should be kept free of dirt and dust. Inspect them often when in use and wipe them down regularly.

• Protect the bulbs from having water splashed on them when in use.

• Inspect each and every lamp several times each day when in use.

• Replace the bulbs on a regular basis to avoid having one fail during a late night or when you are away from home.

• Brooders should be cleaned and disinfected often with a mild to strong chlorine bleach solution or a good commercial product like Oxine. Continue to monitor your brooder units quite closely even as the days begin to warm. Also, the chicks will often let you know when they are in some sort of distress through their loud and excessive peeping.

BROODER MANAGEMENT & CARE

Here are additional items to consider about brooder management and care.

• Yes, upon occasion we've all lined a brooder with newsprint and we've all been lucky when doing it, too. Newsprint is absorbent, but it can become slick and slippery. It can lead to chick losses due to slipping and the resultant problems with spraddled legs.

• Provide a one-gallon waterer for every twenty-five chicks. It should have a narrow or drown-proof lip. I prefer waterers with red bases to better draw the youngsters to the water. With weak and closely bred chicks or bantams you may have to go with the even smaller waterer bases meant for quail.

• Offer the first feed at floor level in a low tray, egg flat or chick box top.

• A sheet of very fine wire mesh in the brooder bottom will give improved footing. It may be placed over the litter material for the first couple of days when spraddled legs are most likely to occur.

• Chicks crowded into the far corners of the brooder unit, with gaping beaks, and cheeping excessively are too warm.

Chicks piled tightly together and peeping loudly beneath the heat source are cold. An old wives' tale from my long ago youth held that you simply did not buy nor hatch chicks to arrive in the month of June. The belief was that those June arrivals would "peep" or "yeep" themselves to death. June is a bit late in the year to have birds ready for the fall laying season, but there are really no grounds for this hoary old tale.

The little chicks excessive peeping behavior can be described thusly. By June the daylight hours in many parts of the country can be quite warm and folks were lulled into the false belief that young fowl could be rushed outdoors quite quickly — too quickly. June nights can still turn quite cool and here June is the month for our "blackberry winter," a handful of quite cool days and nights. On those cool nights the chicks peep loudly to display their discomfort and the chilling can lead to numerous health problems and substantial death loss.

To achieve optimum growth and development, heirloom varieties need to be brooded in a truly sheltered environment. In this situation they are both comfortable, secure and can achieve maximum performance from quality feedstuffs and clean, fresh drinking water.

I think the moment that first marked the sincerity of our involvement in serious poultry production and preservation work was the acquisition of a larger capacity, high quality cabinet incubator. We had worked along with styrofoam units and round, metal tabletop models, but the results we got had been scattershot at best.

Many achieve good success with the smaller units and it is not my intent to disparage them. They are a low cost approach that has a role to play with many small volume poultry producers. They are certainly a good place to begin for those learning about the incubation process.

To say the least, the smaller units can be quirky to use and some people seem to have much better success with them than others do. One regular at our local bird market does the near impossible by regularly achieving high percentage hatches of duck and goose eggs (the latter are always a tough hatch) with a little styrofoam unit. I'm sure he tends the unit quite carefully, but there is still some skill involved to achieve this output.

Most independent producers transition from one or two of the small, tabletop models to one of the cabinet models. They may even continue to use them in tandem or use one of the tabletop units as a hatcher for the cabinet model. Many do not like to hatch in the unit used for primary incubation believing that the wastes and fecal matter produced by the hatchlings make such a unit harder to sanitize. There is also a belief that the dust and debris produced by the hatching process may take a toll on heating elements and moving parts.

The metal units have the longest history with small producers although some in the two hundred-egg range did invest in early kerosene and electric models. A friend has an old redwood model that went from incubator to a rehabbed collectible cocktail table and then back to being an incubator.

The metal models have been around for decades and were made in many styles including a high-topped, "gooser" model for waterfowl eggs. Two different firms generally offered these models. These were Sears, the then rural general store and Brower, a firm with a long history in livestock equipment. The round units are no longer being manufactured, although the Brower Company, now in Iowa, does continue to make replacement parts and they can be accessed on the Internet. A number of the larger hatcheries and poultry supply houses now offer the replacement parts.

These are "still air" units that function best when room temperatures never fluctuate greatly or fall below fifty to fifty-five degrees F. "Still air" simply means that there are no air-circulating fans within the units. An egg turner of sorts was made for some models to rest upon and looked a bit like on old record turntable.

USING "STILL AIR" INCUBATOR UNITS

For best results consider the following operational tips.

• Place the units where temperatures generally move up or down no more than ten degrees or so in a twenty-four hour period. A day to night range between sixty and seventy degrees F. would be ideal.

- Do not place the units in direct sunlight or where drafts can move around them.
- Thoroughly clean and disinfect the unit with a mild bleach solution between uses.
- Have the unit up and running at least forty-eight hours before adding hatching eggs to be sure that the thermostat is running true. Like most incubators they seem to operate best when filled to capacity.
- Replace wafer thermostats at the start of each hatching season.
- For units without a turner, place a lightly penciled "X" on one side of each egg and an "O" on the other.
- Without a mechanical turner, the eggs must be turned by hand. Turn them three times a day by rolling the palm of your hand across the tops of the eggs that are turned on their sides. Turn them from the "X" marked side to the "O" side and then back with the next turn.
- If different groups must be incubated together, also write the starting dates on them with light pencil strokes. Mixing groups is seldom a good idea as the best hatches result when chicken eggs are not turned after the eighteenth day and the unit remains closed until the hatching process is complete.
- Humidity is supplied from a water pan set beneath the mesh floor under the eggs. To improve humidity levels, place a sponge in the water reservoir to wick the moisture upward and around the eggs.
- Monitor temperature with a thermometer positioned to record the cabinet temperature at the egg level. I prefer a digital thermometer with a longer sensor probe.
- The great flaw with these units is that the top must be removed to tend the eggs and water pan. Some get fairly good at tipping it up from one side and working quickly, but this still does cause internal temperatures to fluctuate greatly.
- Hatching can occur from the nineteenth to the twenty-third day of incubation. Some breeds will consistently hatch earlier or later than the twenty-one day average normally given for chicken eggs. Bantam eggs will hatch earlier than standard breed eggs. Cold weather can delay the hatching process and hot weather may speed it up.
- Remove the chicks six to eighteen hours after hatching is completed.

I do not like to see different species mixed in any incubator due to their different hatching times and humidity needs. Waterfowl eggs also seem to bring their own special sanitation needs as they are often produced during the damp and muddy times of the year.

The small, metal incubation units generally hold fifty to one hundred eggs. Most small operations will be able to fill such a unit with eggs that have been laid over a seven-day period or less. The fresher the egg the better its hatchability. Incubators are nearly all bought used now and a friend of mine that deals in used poultry equipment nearly always has two or three on hand. They seem to sell regularly for thirty to sixty dollars each. Most will come missing a thermometer and humidifier pan, but these are fairly easy to improvise and replace. A small light atop the cabinet will tell when the unit is heating. The three most common components apt to need replacing are the thermostat wafer, heating element ring, and the power cord. All can be replaced simply and with few tools. The wafer should be replaced before trying to make any sort of unit diagnosis.

There is something about these units that seems to invite a bit of tinkering and they are a good choice for those needing a dedicated hatcher unit to go along with a small cabinet incubator. They have proven to be very durable units.

The styrofoam box models are the most commonly seen now and I believe this has much to do with the low price for these "bare bones" beginner models. However, they may not really be the best choice for the inexperienced beginner.

The styrofoam boxes come in several cabinet styles and with various sized, plexiglas inspection plates in their tops. They normally hold seventy standard eggs laid on their sides or forty-two when placed in a cabinet with an automatic turner unit. You will see basic, still air models of these units selling new for as little as forty dollars or even a bit less.

If everything breaks right, they will hatch a fair number of eggs, but they require good care and careful positioning. There are a number of extras that can be added to improve performance in these small units and their operational guidelines are the same as for the round, metal tabletop models. The most common extras are an automatic turner, circulating fan, and larger inspection window. The latter is of greatest importance to those using the units to teach embryology and poultry birth.

The automatic turners fit into the cabinet and have their own circuitry. The cabinets will hold forty-two eggs set on end. The turner will make it a bit more difficult to inspect and fill the water reservoir in the bottom of the cabinet. With an automatic turner, the cabinet will have to be opened far fewer times, but the turner unit must be removed for the last three days of incubation or the eggs relocated to a separate hatcher. A barebones still air unit would be a fairly good starting hatcher choice.

A hatcher will cost nearly as much as an incubator unit and will require a separate plug-in electrical supply. The second option to consider is a circulating fan. It will change the nature of the unit and if you can only afford one option this is the one I would recommend.

The fan gives a more consistent and even heat distribution around the eggs. The turner and the fan will each add about forty dollars to the price of the incubator. The incubator itself will come with a fairly simple thermometer to be positioned on the cabinet floor and read through one of the inspection windows. The unit has a water reservoir formed into the bottom, a wafer thermostat, and a series of push buttons on the cabinet to help regulate humidity. A few of the styrofoam incubator models can now be had with an expander strip that will enable them to be used with larger, waterfowl eggs, too. While the fans can be bought after market, I believe that best results (and any warranty benefits if needed) come only with those fans that are factory installed. An incubator unit with all of the extras will now sell for one hundred twenty-five dollars (or a bit more) and to me would make an ideal beginner's poultry raising package.

The styrofoam cabinets are extremely porous and are not easily cleansed and disinfected. Some cleaning and disinfecting products will actually dissolve the styrofoam and rough scrubbing will make them even more porous. Starting to be available now is a stiff plastic bottom liner that can be used with some units and then discarded after a few hatches. Many will simply buy a new cabinet bottom at the start of each hatching season.

A third option is to line the cabinet bottom with heavy aluminum foil prior to each setting. The aluminum foil will mold fairly easily to the contours of the bottom and can be removed after each hatch. Some believe that it may even improve hatching rates.

I have been in homes where six or eight of these units are set up and running with two or three solely dedicated to the hatching process. I have bought several of these secondhand over the years and they seem to pop up all over the place from bird swaps to garage sales to consignment auctions. Look for the best buys in the fall and winter.

These smaller styrofoam units certainly have their poultry raising fans and no doubt, price has much to do with it. With scrupulous sanitation and careful operation, they work with varying degrees of success. The best we have ever achieved was about a fifty percent hatch, but others report hatch rates well into the nineties with them.

As noted earlier, the acquisition of a cabinet model incubator marked both a step up and a step out for our poultry producing efforts. Capacity went up as did hatching rates and we began to expand our hatching times both earlier and later in the year.

Our particular unit holds nearly three hundred standard hen eggs in the three trays and has a non-turning hatching tray at the bottom of the unit. We have it set up atop a small table in an unheated building, and here in the Midwest we are generally able to operate it with good results from late-March through mid-October. With heirloom breeds we normally average seventy-five to eighty-five percent hatches. The up and down swings we sometimes see in barometric pressure here in the Midwest (especially in the Spring) seem to impact the wafer thermostats a bit.

A digital thermostat would add another seventy dollars or so to the five hundred dollar cost of such a unit. The great variety of styles and designs found in cabinet incubators can cost anywhere from five hundred dollars up to several thousand dollars. A company now has even begun to manufacture a kerosene-burning model again. Some are box-shaped, some octagonal, some are domed, and a few have even been made from reworked refrigerator and deep freeze cabinets.

A number of sources are now even offering for sale refitted cabinet incubators from an earlier day. These include old Leahy, Petersime and Ridgeway incubators. Some were old hatchery units or were used in Ag schools and some were home models from back in the day of the larger breeder flocks. Some of these units combine an incubator and hatcher in one cabinet.

A while back my brother-in-law picked up a double unit, 2,500-egg incubator and 2,500-egg hatcher for just two hundred and fifty dollars. It needed a good cleaning and a bit of repair. The real drawback to it though was that the rows of eggs must be turned by hand from simple switches along the outside of the cabinet. A couple of times a day you flip a series of simple levers to reposition the eggs in their racks.

Incubator operation will be discussed in greater detail later, but the real key to incubator selection is to match the unit to the hatching task to be done. There are folks still getting the numbers they need with setting hens and some veteran producers will only trust their most valuable eggs to natural incubation.

With our goal of producing twenty-five to fifty chicks per week for five to six months out of the year, a cabinet model made the most sense. It is a tool of our trade, if you will — one that helps us to do a better job and be more productive with our heirloom birds. With good care the cabinet should have a long productive life and the units seem to retain considerably high resale values.

Buying used poultry equipment can be a good thing if it has been well maintained and not recently exposed to any disease-causing organisms. If you can buy quality equipment and housing inputs for fifty percent (or less) of the new price then they are certainly deserving of your consideration.

BUYING USED POULTRY EQUIPMENT

There are many "buyer beware" tips to consider when buying used poultry equipment. Carefully inspect each item to see its condition and if there is any visible damage or missing parts. If it has an electric cord and they won't let you plug it in — then don't buy it.

Nearly all used livestock equipment is sold as-is and where-is. Occasionally you just need to reattach a hinge, lubricate a moving part or give it a good cleaning. But, sometimes it turns out to be just scrap iron and poor scrap iron at that.

Do not forget to factor in the cost of getting your acquisition home when trying to decide on a price to pay. I was once offered some three-sided, twelve-foot by twenty-foot livestock buildings for free and passed on them because of the costs to get them home.

Sadly, some pieces of poultry equipment now are considered quite collectible and may sell accordingly. Especially pricey are glass and crockery feeders and waterers and any items with elaborate printing and company logos.

Take the time to go and really examine the items and visit with the seller about the reasons for the sale. If the seller says all of his birds just died then pass on the goods. If the seller says all of this "old chicken junk" has been taking up loft space for twenty years then offer to take it off his hands.

Take the time to fully clean everything at the site of initial purchase. Free it from all traces of mud and dried manure before bringing it home.

Once at the homeplace, place the non-electrical items some distance away from any livestock and poultry on the farm. Spray it with a chlorine bleach solution or other disinfecting product. Then leave it exposed to the naturally disinfecting rays of the sun for at least thirty days.

STORING FEED

Poultry feed is generally sold as small pellets, crumbles or meal in fifty-pound paper sacks. Scratch grains are sold either as whole grains or with some of the larger grains slightly rolled and cracked. Digestion aiding grit products are also sold in such bags.

The feedstuffs need to be stored in a way that will keep them fresh, dry, free of insect pests, and palatable. As most heritage producers will be buying these in small lots of a few bags at a time and in different formulations for birds of different ages they will need appropriate storage options.

A simple trash can, with a tight-fitting lid, will hold fifty to one hundred pounds of such feedstuffs. Larger models are also available. A fifty-five gallon metal or plastic drum with lid will hold approximately 350 pounds of poultry feed. Elevate such containers on solidly made pallets or concrete blocks placed inside a dry building. The elevation will prolong the life of the containers and facilitate vermin control around the stored feedstuffs.

For freestanding feed storage that doesn't take up space inside a building we have come to rely upon old, chest-type freezer units that are no longer operable. I incapacitate the locks, elevate the freezers on concrete blocks or treated timbers, and try to position them beneath the overhang of an existing building. One will hold many hundreds of pounds of feed and bags will sit upright in it. Many are often segmented inside.

POULTRY MANAGEMENT ESSENTIALS

As with any business venture it is sometimes the smallest of things that can make the greatest difference. The following are a few of must-have items for poultry raising.

One of the most useful things on the farm is an assortment of all-wire coops. They are used for holding, sorting and transporting birds. Many of ours have slide-out bottom trays that make them especially useful when transporting the birds and displaying them at the farmers' market and other public places.

I use several small cages positioned at eye level to hold birds being evaluated for sale or the culling process. Into other cages I will then deposit those that did and didn't make the cut and relocate them to appropriate holding cages.

At our farm, nearly every scrap of wire mesh is held back for repair needs and future cage building projects. Corners and bottoms are where cages and elevated runs most often fail. The smaller pieces of wire can be trimmed and then clipped into places where existing wire has rusted badly or failed completely.

It helps to keep it sorted into different stacks based on mesh size. For fast emergency patches over larger areas, keep on hand a number of salvaged refrigerator shelves and similar materials to patch the breached area.

In a large toolbox I keep a number of small items including these replacement parts; extra thermostat wafers, humidifier pan sponges, and other miscellaneous replacement parts. Also in the toolbox are often used tools; side cutting (flush cut, if possible) pliers, j-clips and clip rings with pliers for both (offset ring pliers are the most useful), j-clip opener, plastic pull-tites (for fast wire attachment), screws and fender washers (to mount cages), and other miscellaneous tools. To save a great many steps, buy a toolbox large enough to hold all of these.

We use a simple metal break to greatly facilitate cage making and the creation of dropping pans.

A bubble-type egg washer will prove quite practical if selling a number of table eggs.

A long handled dip net will be most useful in recapturing feathered escapees. Likewise a long handled wing and leg crook is good for retrieving birds from distant corners.

Also important to have on hand are the materials for keeping detailed records of flock breeding and performance.

A quality unit for candling eggs is essential. We have a handheld unit that works quite well.

A toe punch and banding supplies are also useful for identification.

Poultry equipment need not be expensive or lavish to produce good birds. As long as the birds are kept comfortable, secure, and well nourished — the equipment and housing is doing its job. Add ruffles and flourishes as you can afford them, but even then only if they will improve bird performance and your job as their steward.

I often thumb through all of those poultry supply catalogs oohing and aahing at all of the latest doodads, but always have to think back to those four Buff Cochin roosters of long ago drinking so contentedly from their tomato cans. Producer imagination can do wonders with even the simplest of materials. I have seen good poultry housing made from old hog houses and even plastic barrels stacked up in a near pyramid fashion. It works just as well to provide the essentials.

POULET ROUGE VS. CORNISH X

Consumers are becoming increasingly concerned with where their food comes from and how it is grown. The consumer responses have driven the growth of open-air and virtual farmers' markets, and increasingly sought direct relationships with their farmers. Unlike conventional or industrial farming, this direct relationship allows transparency and brings animal welfare and husbandry to the forefront. In this extensive (rather than intensive) farming model, farmers cannot "make it up on volume." So selecting breeds that thrive in a natural farming system is of paramount importance. For several decades, the standard meat bird of the American poultry industry has been the Cornish and White Rock cross broilers, commonly referred to as Cornish Cross (Cornish X). These are the ubiquitous chickens with white feathers that have been bred-in so that they leave a cleaner carcass. They were bred to perform (reasonably) well in confinement, to produce the big white breasts that Americans have come to love and to reach live market weights in the range of five pounds in as little as six weeks. But every decision has trade-offs, including the disappearance of breeds that cannot compete with the production rate of the Cornish X. There is also the dilution of genetic variation as the Cornish X takes over as the sole breed of meat chicken. In days gone by, farm chickens would dot the countryside with chromatic diversity, adding not only aesthetic charm, but also contributing a unique culinary experience for the palate of the consumer. Why, do pastured poultry producers so overwhelmingly favor the Cornish X breed? We believe there are four primary reasons for this: 1. They promise an enticing economic return given the efficiency with which they convert feed. 2. They produce large breasts and are the breed with which Americans are most familiar. 3. They are very readily available. 4. It's what everyone else does.

As mentioned earlier, the Cornish X was developed to meet the needs of the industrial poultry model. But that doesn't necessarily make it the right choice for a pastured poultry operation. When we started Nature's Harmony Farm, we were determined to raise animals naturally, and that meant establishing herds and flocks that could do well in our environment. All animals would be rotated intensively across the farm in the elements of nature to improve organic matter and pasture diversity while allowing each animal to enjoy the traits that it had evolved to use. It turns out chickens are quite content on lush pasture.

I suppose this should not have been a surprise to us, but when we first watched the chickens devour grass we were a little surprised. Chickens, be they laying hens or meat chickens, are a definite asset if your goal is to improve pasture and soil quality.

With taking flavor back in mind, we were very skeptical about raising Cornish X chickens. We had read and heard so many horror stories about how they would grow so fast that their legs would be unable to support their massive weight, and they couldn't get around. We viewed them rather as genetic freaks. In looking for alternatives to the Cornish X, we were intrigued by the Label Rouge program in France. This program began in 1965 as an initiative by small farmers and poultry breeders concerned with maintaining consumer confidence by improving and guaranteeing animal husbandry methods. Label Rouge is not a single, stand-alone brand. Rather, it's a rigid set of standards adhered to that promises quality to consumers. Even though Label Rouge poultry is roughly twice the price of conventional poultry, today it commands over 30 percent of the entire market in France. This is a remarkable success story that demonstrates how much consumer's value, and are willing to pay for, quality. Label Rouge does not include a single breed of chicken, but only certain slow growing genetics are allowed that are well suited for outdoor production. Various "farm" breeds may be used, and all must be raised in accordance with strict standards, which include:

- All birds must be raised a minimum of 81 days.

- All birds must have access to the outdoors from 9 a.m. until after dusk, and birds must be outside for at least 42 days of grow-out.

- Feed rations must consist of at least 75 percent cereal, may not contain animal products or growth stimulants, and must be non-medicated.

- Birds may not travel more than two hours to processing, and all birds must be sold fresh within nine days after slaughter.

We felt aligned with these values, and sought out breeds that could mimic this approach in Georgia. We do not replicate the approach precisely — after all, what's right for France is not necessarily right for Georgia — but we like preserving breed diversity and we liked the longer lives that were afforded the slower-growing birds. Our search for chickens that were similar to those raised in France led us to two breeders who actually had imported their breeding stock from birds used in the French Label Rouge program. Excited, we ordered multiple batches from each hatchery, starting with 200 in March. We followed this up with an order of 250 from the other hatchery, and subsequently ordered almost 500 more. All told, we raised about 1,000 of what we call "Poulet Rouge" chickens. The name is a bit of a misnomer, in that they are not all red — but they *are* colorful and beautiful to look at! We raised our Poulet Rouge chickens in portable chicken tractors that we moved to fresh pasture daily. For the first batch, we actually allowed them to free range, only locking them up at night. For the second batch, we kept 50 or so confined to each pen to study the differences in mortality, growth and performance between free-range and daily move systems. All of the pullets were purchased as day-old birds that we brooded on-farm. Now back to the Cornish X and despite all the perceived negatives — we wanted to know for ourselves. So, late in August, we ordered what would be our first of two batches of Cornish X. The second batch came in mid-October, so we could evaluate the Cornish X during three seasons of the year. What did we learn by raising both Poulet Rouge and Cornish X chickens? Our two main subjective findings are: 1. Poulet Rouge meat is definitely more flavorful. 2. There are no more "net" problems raising the Cornish X than there are raising the Poulet Rouge. The problems are just *different*. The diversity of these breeds far surpassed what we could have anticipated.

These birds are different in almost every respect. Here are our key findings in more detail:

- Cornish X eat more from day one and quickly fall into the routine of sitting by the feeder. When it's feeding time, they attack your feet as you enter the house in anticipation, whereas Poulet Rouge tend to be a little less gregarious.

- On pasture, Poulet Rouge spend more time standing or roaming, while Cornish X sits. The Cornish X is not a good forager, for either insects or grass. Conversely, the Poulet Rouge is a very strong forager for both.

- The growth rates are very different. Cornish X grows before your eyes. Cornish X is ready for processing in no more than eight weeks. The Poulet Rouge takes 12-16 weeks to reach a similar size.

- We experimented with letting both breeds go well beyond their normal lifespan as a meat bird. The result is that the Poulet Rouge are much more active as they get older. The Cornish X slows down as it gets older/bigger. Older Poulets that are large still walk and move with ease, whereas Cornish X waddle and take more effort to get up.

- The Poulet in general fight more and will peck each other at times. This occurs even in a complete free-range system where there are no space issues.

- While we experienced some minor predation problems when the Poulet Rouge free-ranged, we would not feel comfortable raising the Cornish X in a free-range system. Poulet Rouge is more alert and more able to flee. Cornish X is much more vulnerable.

- During processing, the Cornish X is easier to pluck, due to the white feathers. The dark feathers of the Poulet Rouge leave a carcass that is not as pristine as most consumers have come to expect.

- Poulet Rouge meat is firmer, which may seem stringier to some. The meat is also juicier. Cornish X meat is drier.

- The Poulet Rouge performs similarly regardless of the time of year and weather. The Cornish X does not. Very hot weather and very cold weather are not a good match. Cornish X was bred for confinement and climate control.

- Poulet Rouge meat definitely has more flavor. It is cleaner tasting, and tastes more like what most people think chicken should taste like. The Cornish X has less taste.

- The Cornish X has a much larger breast and more white meat. The Poulet Rouge has a longer, flatter breast and more dark meat. If you're looking for that Chick-fil-A piece of chicken, that's the Cornish X.

So, where does all this leave us? For starters, we will raise Poulet Rouge chickens for our own consumption if for no other reason. The taste is too good not to. And we have developed a market for them in north Georgia, as educated consumers and those with a passion for cooking seek to introduce new flavors to their taste buds. We were bombarded with questions on how to best cook them, but those who have bought them are eager to buy more. Finally, it should be no surprise that the majority of people just want clean, safe chicken at an affordable price. So we will continue to experiment with Cornish X at times of the year that are propitious so that we can serve both markets. In the end, nature and our customers will pick the breed that is best suited to our environment.

Tim & Liz Young raise grass-fed lamb and beef, free-foraging pork, pastured poultry and heritage turkeys at Nature's Harmony Farm in Elberton, Georgia. For more information, visit www. naturesharmonyfarm.com.

Originally published in Acres U.S.A.

· CHAPTER 5 ·

BUILDING THE FLOCK — SOME ASSEMBLY REQUIRED

LIGHT BRAHMA
• MALE •

LIGHT BRAHMA
• FEMALE •

SILVER-GRAY
• DORKING MALE •

SILVER-GRAY
• DORKING FEMALE •

WHETHER IT IS YOUR GOAL TO HAVE FIVE, FIFTY OR EVEN FIVE HUNDRED BIRDS, THE TASK INVOLVES THE VERY SERIOUS BUSINESS OF BUILDING A CHICKEN FLOCK AND YES, A CHICKEN BUSINESS. Very rare breeds generally have to be started quite small and the flock-building task may be a work of many years. With the very rare, the best to be hoped for is to begin with but a handful of birds from which to build upon. Many very rare breeds may well be lacking in type too. When there are just a few to access, then that is where you have to make a start.

Even with a more mainstream breed, from a box of twenty-five as hatched chicks about the best that can be hoped for is one good trio and one or two other, usable pairs. The ideal would be to begin with trios or pairs of adult or started birds from two or three different and distinct flocks — or plugged in quite closely to one good breeder's existing program. If not using developed birds, then start with as large of a lot of chicks as possible from a similar number of distinct sources. Either way there will be two or three sets of birds from established lines with which to work. Then birds bred from these sets can begin to be combined to create your own breeding line or lines.

A certain number of small-scale producers will inevitably continue with the traditional practice of restocking commercial laying flocks or meat bird groups from distant off-farm sources. These producers should be looking for sources of supply closer to home and will be best served by other, independent producers. Some are beginning to envision poultry producers again being served by poultry breeders based no more than an hour or two away, thus eliminating dependence upon mail deliveries and far distant sources of supply.

With breed choice or choices made, markets targeted, numbers decided to grow to set, and initial seedstock on hand, it is time to set the breeding plan into motion.

When beginning with chicks from a hatchery or other single source it is best to assume that they are all going to be full or half-siblings. This is a tough place from which to make a start. The brother x sister mating is genetically the closest of all. Still, when it is the only option, just swallow hard and do it. It is however, a situation that can be worked through with an appropriate plan of action. Prolonged very close inbreeding can impair vigor, reduce fertility, and increase the risk of certain defects. Some of these can be minor and some potentially quite lethal. With rare breeds coming from small populations, impaired vigor and reduced fertility are the most likely to be encountered. To that end always put a strong emphasis on growth and vigor in the selection process. Call it the will to live or the ability to thrive, but it has to be there ahead of anything else.

Pull your keeper pullets only from the largest and most vigorous one-third of their hatch group. Males should come only from the top one-fourth. The breeding program will be boosted farther and faster if even more strictness is shown with selections, but it will take longer to build numbers.

Don't forget, just as inbreeding without good structure can pull bad traits to the forefront, well planned inbreeding (linebreeding, actually) enables producers to key on the exceptional traits of even a scant handful of good birds. Many of the best flocks and herds in the United States often trace back to but a single individual of great merit and often that one individual is female.

If building from a single source and with limited knowledge of those genetics breeders must cull ruthlessly for hardiness, vigor and type. This must be continued with all of the generations to come too. With some of our rarer breeds I deliberately look for males with some real "rattle and snap" to them. To me this is a really developed libido along with good size and substance. My hands and arms show some of the evidence of their gingery attitude too. I don't want a male to be a pit-ready battle stag, but I want him up for his hens, on his toes when approached, and showing his stuff with his wing down and sweeping.

Early in the spring of 2006 raccoons hit us hard. A large pair took several valuable birds. In one pen we had a set of Rosecomb Rhode Island Reds headed up by a two-year-old male of good size. He had lost one hen and was a bloody mess when I found him, but the pen next to his had been totally wiped out. With much of his comb gone he took a couple of weeks to heal and he will never be pretty again, but he continued to fill eggs for the whole season. He has earned a place in our breeding pens for a long time to come with that kind of grit. To this end, in the 2007 season he was penned with some of his best daughters in what many call a "rolling mating" program.

With a first generation hatched from single source birds, the breeding options begin to open up a bit more. One to consider practicing is "rolling mating." Take the best pullets produced and breed them to the best males of the first generation. Select the best male from the first chick crop and breed him to the best females of the parent generation. Parent x offspring matings aren't nearly as close as sibling matings and many flocks have shown good results by continuing with the rolling mating for generations. With this program two separate breeding pens must be maintained and all retained birds must still be carefully selected for size and vigor. Also, always be on the alert for any of the performance-robbing traits indicative of a mating tilting too far. Watch closely for slow hatching, reduced hatches, crooked toes, crossed beaks, and the like. Not every crooked toe or crossed beak is from a too tight mating and many things can influence a hatch, but do consider these as potential red flags as the mating program unfolds.

Better line mating choices are cousin to cousin, aunt to nephew, and uncle to niece. As numbers build, such matings become more easily executed. Some outside breeders can even supply an initial round of birds that are so related. This will lock a producer into that breeder's program, but if it is a good one, it will move the breeding that much closer to your goals.

The purpose of linebreeding is to create a set of birds consistently strong in a number of desirable traits. Then these birds can be bred regularly with a strong prepotency for these traits. It is all about building a very high degree of predictability into each new generation of offspring. A heavy amount of inbreeding went into the creation of each and every pure breed and when done correctly it is a most useful tool for all livestock breeders. Linebreeding will find any bad traits in a line, no matter how deeply buried, but it is also the best way to build upon and enhance the good traits. It is up to the producer to keep records, document birds, and monitor their individual performance closely enough to be assured that all is in good order and that there is no fall off in economic or reproductive performance. To borrow an idea from the old Kenny Rogers song, the key is to know and decide quickly which ones to hold (and use) and which ones to turn into chicken salad.

Another method to consider, when all birds in the flock are identified and marked as to their breeding, is to establish "breeding families" within the flock. Some families will descend down from certain top females and others from exceptional males. The best male from a male line can then be bred to the best females from a different family. All have a similar background, but males and females from the same family are never mated together. This can be rather complex and entail a lot of bookkeeping and small breeding pens, but it is another proven breeding method.

With started or adult birds many months of breeding time, if not a full year or more, are gained. The result is actual made birds, which give the clearest possible picture of what they are and what to expect from them as potential breeders.

With older birds from breeder sources buyers will generally receive much more and much deeper documentation as to their breeding. With such birds from two or three different sources producers can then both breed onward from those lines and begin combining genetic pieces from the different lines. All the while assessing all lines as to how they perform in their new environment on the small farm or poultry yard.

Matings of birds from different strains from within the same breed are said to be "outcrosses." Commercial producers sometimes use an outcross male to ramp up performance for an economic trait like meat yield or egg production, if the male comes from a line very strong or prepotent for such a trait. An outcross to a closely lined flock can really open up things genetically and will be a mating with far less predictability.

Some type factors may emerge quite differently following such an outcross. However, an outside male can blow both feather and breed type as it relates to conformation. Good breed type and performance traits can be melded together into one flock, but it will demand some well thought out and scrupulous breeding practices. Build carefully on strong birds. A Rhode Island Red flock with a laying average of one hundred sixty to one hundred eighty eggs per hen per year is a reasonable long-term goal. With this flock, some females may even crowd the two hundred egg figure. Over time, growth rate and meat yield can also be likewise improved from breeds like the Rocks, Wyandottes or the Delaware.

Regardless of species, if you wish to open a debate among livestock raisers simply ask them how they plan a mating. There is sometimes a certain serendipity to things genetic. Some hold that if birds of the same pure breed are mated together they will reproduce to the average or mean for the breed. More likely it will be for the average of the flocks from which they are sprung.

Those that have worked with poultry for very long have no doubt heard or even experienced something of the following bit of anecdotal evidence. As breeding seasons wind down some producers start putting all of the birds of a breed into a single, larger flock to simplify choring. Eggs from the essentially random matings that occur within this larger group then will go on to produce some of the best chicks hatched that season.

Still, the two most common types of planned matings are simply termed "breeding best-to-best" and "breeding strengths-to-weaknesses." Each is based on carefully and honestly seeing the birds before you, comparing them to that ideal type picture that is carried in the mind's eye, and then factoring in how the birds have been performing in their role as egg layers or poultry meat producers. That honesty of vision is critical, as we all seem to succumb to at least some degree of "hen house blindness" from time to time. At one time neighbors or even Extension Agents were invited in to assist with such evaluations and flock cullings. It still pays to call upon a more impartial second pair of eyes to evaluate birds whenever possible. The guest judges should however be mindful of your goals and how the flock is managed. Breeding "best-to-best" is just that. The very best male in the flock is bred to the very best females. Some will do this regardless of how they may be related. The goal is to put a double dose of those good traits into the resulting offspring. Those traits that aren't upgraded should at least hold to an already fairly high standard, in addition to always actively selecting and breeding for size and vigor.

It's not foolproof, but getting close enough will produce a flock that is building on some quite strong genetic constants. I have a good friend that has had linebred Hereford cattle for decades and will not hesitate to make a best-to-best mating from anywhere within his linebred herd. He shies a bit from full sib matings, but even half-sibling matings get regularly made there.

Strength-to-weakness matings are essentially fault correcting or performance enhancing matings. It is how we operated our purebred swine operation for well over thirty years. It is a process that begins by being a good student of your breed or breeds and an even better student of your own breeding stock and their offspring. Ours was a propagating herd producing breeding males and females for commercial producers.

This breeding process begins by identifying the greatest flaw or shortcomings in your breeders and/or their offspring. Then select a male with greater strength in that area and begin breeding him into the body of the flock. This should be done slowly to be sure that the mating works or nicks. The bird should be strong in all other pertinent traits including breed type, but if your existing birds are good in certain traits then you may be a little forgiving of one or two minor deficiencies to gain a real boosting in the problem area.

In very closely linebred flocks, a boost can be made from an outside source, but must be managed very carefully. It generally is brought in with a very good female who is used slowly and carefully in side matings before any of her genetics are introduced into the body of the flock. It should be noted here that it is the male side of the mating that will actually have the greatest influence on the female offspring and the hen side that has the greatest influence on future males. The rooster carries the traits for such needed variables as eggshell color and intensity and good layer body type. The female has great influence in the matters of color and feathering. The real influence of a bird may not be felt until the second-generation offspring start making their way into the flock.

In breeding for trait or performance upgrades it is always tempting to seek out some sort of super individual to try to give a great many traits a jump start with a single mating. It is an idea that is asking for too much and it is actually a very unbalanced way to build a mating. With this, as many or more desired traits may slip away as are enhanced. In a strength-to-weakness mating, focus on just one and no more than two traits. When doing this move very slowly and carefully. It will not be a success unless this new trait strength becomes solidly fixed. Always try to retain as much balanced, middle-of-the-road genetic type as possible in the birds being bred.

With all of the above said, it must still be conceded that when making a start with heirloom and rare breeds, it is only the purebred birds that are available that can be used and this is regardless of the state in which they are found. I have owned and worked with my fair share of birds with freeze damaged combs, missing eyes, and far less than showroom type.

Mismarked birds, birds with size issues, birds a bit past their prime, and even some with defects may have to be pressed into a preservation-breeding program in its early stages. Get a current copy of the *Standard of Perfection* and learn by heart what constitute defects and disqualifications for your chosen breed or breeds.

Winning showroom type, especially if it is fad driven and at odds with breed standards, may not even be the goal if producers are striving for a more performance-based flock trait. Still, always strive to maintain good breed character in birds.

Our Exchequer Leghorns, for example, present three breeding challenges at the moment. They could be bigger, they are carrying too much black coloring, and we get too many with pale, off-colored legs. I would like to see ours grow a bit quicker too, but each problem should be addressed in good order. With the above in order we will then be working fairly continuously on egg size and egg production per bird. With every flock there will always be something that needs addressing and that is true of flocks that have been in existence both for one year or twenty years.

Where numbers are extremely tight, birds are too tightly inbred or there are some real type needs, producers may have to really look outside of the genetic "box." When a handful of White Houdans were discovered a few years ago, the need for revitalizing and type improvement were critical and most of the birds had some age on them. A Mottled Houdan hen or two of very good type were selected to be bred with these birds to boost them and their genetic constitution. Some color issues emerged, but the birds are well on their way back to genetically pure status.

Here is how that genetic restoration works. For example, with a White Rock hen and a New Hampshire rooster, it is possible to gradually breed toward pure New Hampshire breed status. Begin with birds that are similar in size, conformation, skin and leg color, comb type, and the like. From the above cross, breed the pullets that are most vigorous and closest to the New Hampshire in appearance back to their sire or another purebred New Hampshire male of good type. The result will be three-quarter New Hampshire birds in just the first generation.

By the sixth year breeding each new generation back to purebred birds and the offspring will be 98.4575 percent pure New Hampshire Red. At the eighth mating back to pure they will be 99.5 percent pure New Hampshire breeding and can be safely considered and marketed as purebreds.

Breeding out to a standard Black Breasted Red Game female with correct leg color (very critical) may be an even better outcross for invigorating some long neglected birds. The black-breasted red color pattern is the base or wild color pattern from which all domestic chickens are said to spring. It is the color all breeds will revert to in fairly short order if left to breed on their own. With a plan to restore a black-feathered breed it may be best to select a black Game showing no purple sheen in the feathering and with appropriate leg color.

From the Game hen you will draw the justly celebrated vigor and hardiness of her ilk. These are factors often sadly diminished in populations that have dwindled down or been allowed to fade. Never breed white into buff either.

With some of the truly endangered and lost breeds some efforts have been made to recreate them. The old literature often contains detailed accounts of the breeds and breeding combinations that went into the creation of a great number of breeds and varieties. A friend has begun assembling the genetic pieces to recreate the Lamona although there are recent reports of a population of this breed having recently been discovered. White Chantecler populations were recreated both here and in Canada, their country of origin, after the breed had been believed extinct. A number of small groups did emerge and we even see this breed at our local swap now. It is even bred now in buff and partridge varieties. Even new breeds built from others such as the Braggs Mountain Buff are beginning to appear.

The guidelines are out there yet for everything from Lavender Laced Wyandottes to Buff Barred Plymouth Rocks. There are those about the task of erecting new color varieties within breeds or bantam versions of long established standard varieties. Here in our area I have seen true breeding populations of Blue Dominiques and Blue La Fleche. Neither one is sanctioned yet by the American Poultry Association, but both do illustrate what can be still done with the heirloom breeds.

As time passes numbers should build and breeders can become ever more discerning in the culling process. One important task even in the earliest building stages is to get some of your production distributed around as a protective measure and to advance the cause of their preservation.

Breeding partners can help with preservation work in a great number of ways. At the most basic level they get numbers widely enough dispersed to assure that at least a portion of the population will survive a catastrophic event. A single predator attack, windstorm or localized power failure can spell absolute ruin for a small, too closely held population. Two or more cooperating producers working together can share expenses and have multiple input into any plan of bird preservation.

The more environmental challenges that can be presented to a line as it develops — the better. If a line is dispersed across an array of environments its genetic code should emerge differently in the face of each different challenge. Thus will the population be given a broader base and one with its gene base strengthened. Down the line, even the best breeding program may show come evidence of unraveling at least slightly. Fertility may fall off, eggs become slow to hatch, chicks emerge weak at hatching or they may even die in the shell.

POULTRY BREEDING IN A NON-STERILE ENVIRONMENT

I've been thinking a lot lately about breeding and it's helped me to flush out some ideas and draw some conclusions that seem to be far against the mainstream. Maybe I'm the one who is way off. After all, I'm still new at all of this. But maybe, just maybe, I'm on to something.

We've been dealing with some sinus trouble in our layer flock. We had a healthy flock of laying hens, but we decided to add some new point-of-lay hens from a nearby breeder to supplement our flock. Within a week, our original hens got sick. Egg production dropped dramatically and the hens began to sneeze and get swollen sinuses. Coincidence or did the new hens bring in something? They have showed signs of this cold off and on for the past 3 months. We talked to the local poultry lab and they said that it could be a mycoplasma infection, which is very common, spread by the air or on people's clothing, but that the birds will be carriers for life and they recommended destroying the flock and starting over and adopting strict bio security measures where no one visits the farm ever. This was a shock to me! I can't do that! Just because they are sick? There had to be a better way, so I began investigating. It lead me to learn that if it were a mycoplasma infection, then symptoms will appear during a time of stress, but otherwise the birds can live and be productive and this infection is of no danger to anyone else (it's basically like a bad and recurring sinus infection). So I figure, let's not stress out the birds and let's boost their immune system so that they can keep symptoms at bay. We've been giving them apple cider vinegar and garlic in their water any time the symptoms begin to reappear (usually after a cold or wet spell). Temps in the 60's for the past 4 days have really helped them all to bounce back and they look great now. But I understand, that they may be carriers and would pass this onto their young if they were to breed.

We considered not breeding them, but after careful deliberation we decided that breeding this flock was absolutely the right thing for us to do. During this flock's bout with sinus trouble, we were also raising a batch of day-old chicks in the brood house to be our future second Eggmobile. These birds are now free-ranging and have not shown any symptoms of sinus trouble even though the lab told me that this infection can travel a 1/4 mile in the air so soon all of our birds will get it. After reading about a vaccine for mycoplasma I learned that it was a live culture vaccine, so they would actually be giving the young chicks a small dose of the live bacteria so that their bodies would fight it off and therefore build up some immunity to it. Well, if this infection is in the air, then our day-old chicks were inadvertently getting a small dose of it. I believe they are healthy now because they have built up an immunity. We also recently bought some more 4 month old point-of-lay hens from a self-contained poultry house and sure enough, they got the sinus trouble within a week — no immunity.

So, when I speak to the experts they recommend a sterile environment — vaccinate and give antibiotics to erradicate the bacteria. Then raise the chickens in a house where they don't come into contact with anything. I say that a sterile environment will weaken the immune system since the birds never have to fight anything off and therefore never build up antibodies to proctect themselves. Imagine what would happen to Bubble Boy if you let him out of the bubble. We have a natural farm — not a poultry house — so our animals come into contact with anything in the air, wild birds, and lots of visitors. This will never be a sterile environment. What I've learned is that the strongest survive and if we breed chickens that have overcome and learned to live with mycoplasma or whatever it is they have, then they and their chicks will be stronger in our particular environment. I've also learned that if we bring in birds that are not adjusted to our environment and that do come from a sterile house, then they can have trouble, which emphasizes our goal of having a completely closed flock where we produce our own chicks and never have to rely on hatcheries or other breeders.

I remember my first year of teaching elementary school. I was warned that this would be the sickest time of my life, but that within 3 years I would be impenetrable to all of the common colds, flus, ear infections, strep throat and whatever else those germy little kids carried into my classroom. Sure enough, they were totally right. At first, the introduction of new bacteria to my system made me sick, but I fought it off, and then I continued to come into contact with low doses of these illnesses, which in turn just boosted my immunity to them. By the third year I was never sick, despite the many times that I wiped runny noses, got sneezed and coughed on, or received hugs from some cute little kid with a fever.

The poultry lab is currently testing some blood samples to see what our chickens actually have, but we've decided no matter what it is, let it be. Let nature work it's magic and let survival of the fittest reign. Hopefully moving towards a closed-flock breeding program will send us onto a path of birds that are made for our environment. I picture a farm full of Super Birds — able to withstand the harshness of the weather, fight off common infections, and avoid the dreaded hawk!

Posted by Liz Young on the Nature's Harmony Farm website at www.naturesharmonyfarm.com.

Once the need for some new blood is clearly indicated, how is it best to add it? How can this be done without derailing a long time breeding program? For many the first thought would be to bring a new male in from outside, breed him to as many hens as possible, and get those hatching eggs filled as quickly as possible. There goes genetic predictability and if he doesn't nick, the whole thing is back to square one or maybe even worse.

Opt instead for a single female of exceptional merit and if she is a bit distantly related to your birds — so much the better. A lot of my purchases over the years have been to get my stuff back. She should be of good size and also display exceptional vigor. This hen should then be mated to your very best mature rooster, a bird from right out of the heart of the flock. Should this mating produce birds of the desired type and vigor, select a handful of the very best females from the mating. Some will then go so far as to sell the hen to remove her completely from the established gene base. Then, very gradually introduce them into the existing breeding pen or pens. Monitor the matings involving these birds quite closely and don't hesitate to pull the pin on them if results begin to falter or matings fail to produce the desired results.

The thing to remember in any sort of breeding program is that you may have to make several starts before achieving the results that you desire. Give a set of birds the time and opportunity to show what they are capable of, but don't hesitate to fold the whole lot of them if they don't then measure up.

Good ones don't just happen and some breeds may take many, many months to develop fully. A noted Black Java breeder once made the point that beyond those birds with obvious physical defects he does very little in the way of culling before the birds are twelve months of age. It simply takes that long for those large frames to develop and the rest of the bird to fill in to match the frame. There are also color patterns that take time to develop and fill.

The industrialized, commercial strains have many believing that a chicken now is a finished product at between six and twenty-two weeks of age depending upon which commercial use that it is being raised for. The heritage breeds were meant to be in place about the family farm for far more than the five hundred eighteen days a commercial layer could now consider her normal lifespan. Heritage breed hens may be part of a breeding and laying flock for four, five or even more years.

Very rare breeds don't remain rare for long if they achieve any level of performance and acceptance. They can reproduce almost like feathered rabbits and a good trio can become a hundred or more birds in just their first season of production. How many of that one hundred will be keepers will still have to be determined.

The exhibition breeder may hatch eggs from just a handful of birds and for only six to eight weeks early in the year — to have a show string in good feather for the show season later in the year. They may then set another small round of eggs later in the year if they need breeder replacements. Most hatcheries and seedstock suppliers run their incubators from mid-February to mid-June or a bit later before breaking up their breeding pens and even liquidating some flocks.

Lots of birds can enter the world in the course of a single year and thus we see once limited breeds like the Dominique now becoming quite plentiful. So plentiful are they now that most catalogs list them. Not every one sold though should go on to enter a breeding flock. Through the years a relative handful of breeds held up their numbers and even with those some real quality and production issues come to pass. A few became even a bit mongrelized — such as the hatchery run Ameraucanas that breeders of that variety now dismiss as "Easter egg" chickens.

Into the 1960s there still existed a fair number of hatcheries and independent breeders with true performance strains of a number of different breeds. I can well recall one Iowa hatchery that each year devoted several pages of catalog space to its particular strain of "Danish" Brown Leghorns.

Other hatcheries and breeders had noted lines of White Leghorns, White and Barred Rocks, Australorps, and a great many more. They didn't just promote what they had, but worked most diligently through selective breeding to make them ever better and more productive. I know several who earnestly search for any remaining traces of those performance bred birds and there are a handful of producers continuing this tradition.

Selective breeding for improved performance hinges on knowing each bird in the flock and how their offspring are performing. Every male and female must be accountable. With small numbers, keep track of each bird with toe punches, leg bands, wing bands or a combination of these. Match these identification tags with small pen matings and very detailed record keeping. It is important to know where every rooster you own is at every waking minute. At one time trap nesting was used widely to track the exact breeding behind each and every egg produced. It was true pedigreed mating.

The hen, with a numbered leg band, would enter the nest to lay and would not be freed until her leg band number was recorded on the egg she had just laid and in the producer's daily log. The male to which she was exposed would also be documented. The eggs then went into the incubator separated by their matings. The chicks would then be immediately toe punched as to identity as they were removed from the hatcher. There is much more on incubator management later in this book. With small numbers there are some similar measures that can be followed that will yield nearly every bit as exacting results. Small pen matings involving just one or two females and a single male can be set up. Full sibling males can be used to breed larger groups and single pair pedigree side matings can all be made with relative simplicity.

The case for knowing and tracking each individual bird in the breeding flock cannot be made strongly enough. It is held that females will hold the greatest influence over offspring size and breed type. The male will influence the finer points of color including undercolor and many of the numerous finer points that go into head structure. Each year producers may have to make multiple side matings to trial new blood, address subtle faults, and continue to tweak flock performance for economic traits. These must be done most carefully and in very controlled situations. Sometimes breeders will even use a bird rather like a dash of pepper on a steak.

If you need just a trace of that breeding, it needs to be applied to only one area of the flock, or overuse can have most serious consequences. It may be found that two individuals work well only when mated to each other. When a mating that works truly well is determined it can be made repeatedly for a goodly number of years.

Dotted around our poultry yard each year are a number of small pens and even all-wire rabbit cages suspended in larger buildings. Into these will go pairs and trios, new birds being trialed and the odd color or other breeding experiment or two.

From a small breeding pen of three females or less and one male, it is fairly easy to determine the per hen laying rate. Coupled with basic laying hen culling skills, it can be fairly simple to determine which hens are laying well and which aren't laying well.

A few years ago I was sent a small group of Buff Plymouth Rock breeding hens that a friend dropped off here at the house. It was well into the fall and the birds were into the latter stages of their laying cycle. Three of the four two year-old hens were clearly showing evidence of having had a long season of laying.

However, one hen was still in full feather, was in better flesh, had a bright red comb and wattles, and her legs were still a bright, waxy yellow. My friend delivering the birds from their original owner remarked that it would be easy to nick out the best bird in the group and nodded at that particular hen.

Actually, she was the least productive of the lot as all that she was eating just went to keeping her looking "catalog pretty." A year of laying and the demands that it makes on a hen's body were manifested in the others in the form of faded and shrunken combs and wattles, legs grown pale, and the overall signs of wear from a season in the breeding pen. When taken in hand for a more detailed examination, their vents were still a couple of fingers wide and moist in appearance — they still had some more lay in them for that season. The showy birds' vent was only about a finger wide and rather drawn and dry in appearance.

Laying and breeding flocks can all benefit from regular, close inspection of the hens and the males. Remove birds that appear ill, are unthrifty in appearance, and are showing signs of low levels of productivity. In the early stages of flock development producers may have to cull rather substantial numbers, but when only breeding from only the best, quality should improve all across the flock.

The small breeding pen system has worked the best for us and with a few pencil strokes on the eggshells to track the matings; they are fairly simple to track from lay to hatching. I have to admit that when numbers got too great, some of our small pens did not get as much of our focus as they should have and our favorites perhaps too much attention. All of us have only so many hours in the day to give to any project, so match the flock size very carefully to the available labor to tend them.

In the breeding pens, keep everything fully documented by using field notebooks, incubator logs, and then a permanent record where the day's field data can be transferred for even better safekeeping. The beginner may feel most comfortable working with just one or two breeds. There are some producers that spend a full lifetime tending but a single breed.

A part of building the smallholding poultry flock is the decision about what the ultimate number the flock is to be. The farm enterprise mix may be skewed too far in one direction if much beyond three hundred to five hundred laying hens are kept. On a well-diversified farm, the individual enterprises complement each other and do not compete directly for the most apt to be in limited supply factors such as space, capital, and producer labor. At about eighty percent production, five hundred hens will produce nearly two hundred fifty dozen eggs per week and the challenge will become both tending the larger flock and finding the time to direct market their production to greatest financial advantage. Farmers can't be lambing, working in the market garden, cleaning the chicken house, and selling eggs at the local farmers' market all in the same Saturday morning.

With the above example and figuring two-dozen eggs per family per week you will have to access one hundred and twenty-five families a week to market the eggs from five hundred hens. A customer base of twice that number of families may be needed to consistently sell that number of eggs week in and week out.

Fifty heavy hens at eighty percent production and with eighty percent hatchability will produce 225 broiler chicks week in and week out in prime hatching season. This is nearly a thousand birds a month to get sold and will require a lot of incubator capacity. In both of these examples, the numbers are not great, but in most instances are at odds with the concept of modern, niche production. The successful producer first finds that niche and then grows only to its capacity to consume at truly profitable levels and then expands no larger.

In late fall, with only a few females in production, our refrigerator still fills quite quickly with surplus eggs. The same is true again in early spring before we turn on the incubator. A big part of the dollars received for farm goods results from the control you have over the quantity of your product. Excessive production will find a price, but too often it is at less then the cost to produce. Over produce and you may well compromise and possibly even destroy a small niche market that has been paying quite well.

We live in a time when the common thinking is that a "flock" of chickens must number into the tens of thousands to have any sort of credibility. While the commercial sector keys on earning power from a house of broilers or layers, the smallholder building numbers with the heritage breeds should key on what is done with each and every bird.

With our micro-hatchery we have quickly learned that each breed has a certain level of demand and a selling price at which it will find a home. Here Buff Orpington chicks outsell just about anything else in buff down and their selling price, early in the hatching season, is going to be up or down a dime (or two) from a benchmark two dollars each. As the hatching season progresses prices tend to go down. Producers can add to a basic selling price for enhanced bird quality, but that, too, has some limits. Car salesmen sell way more Chevy's than Cadillac's don't they?

Presently, for our market the demand for top end Buff Orpingtons can be met with as few as five good hens (in a one cock pen). Ten to twelve hens in production is our goal for a few years from now and if things progress as hoped, twenty-five good females would be as large as we could possibly grow with this breed.

Few are the breeds we have offered that we felt the need to have more then five to eight of each type of hens laying. In our first year with Welsummers, our trio and then pair actually covered just about everything that we could get sold at our local farmers' market. With that breed back then we were actually ahead of the curve locally and had to do a lot of educating to generate sales. Never assume that everyone knows what you know and likes what you like.

Yes, it would have been very nice to have had twenty to thirty Penedesenca hens in production as the demand for that breed began to build. It might even have resulted in being the largest producer of them in the nation. Indeed, there were those who paid substantially for even a handful of chicks or hatching eggs when this breed first started catching national press. Still, Penedesenca could have failed to launch or numbers may continue to grow until they become quite as common as White Rocks.

I must also recognize that my numbers won't crunch the same for other, modest sized producers. Some family farms could and should grow to the maximum allowable level of twenty thousand, directly marketed broilers yearly. Conversely a thousand bird flock selling seventy dozen table eggs daily for $1.50 per dozen won't be as profitable as a three hundred bird flock selling twenty dozen eggs a day for $2.75 a dozen. Great numbers far more often weigh down markets than they build producer profits.

I have written this often over the years about nearly all livestock species (and it will ever be true) — the small producer's commitment must always be to quality over quantity.

While good breeding flocks can come from quite humble beginnings and modest numbers — at their base must be one or two birds of exceptional merit. One good bird is worth far more then one hundred or even one thousand of the average or poor kinds in the flock building process. Before trying to produce as many as possible, be sure that you are producing the best that you can.

To know if they are good, you must know which is which and what each and every breed is capable of producing. Alas, with chickens there are no protruding ears to notch nor sufficient neck strength to support neck chains with tags. To identify individual birds you must look lower down on the body — to the wings or legs and feet to be exact.

It is possible to clip to the fleshy part of the bird's wing what is termed a wing band. To be clearly read the birds must be in hand, but this is a fairly dependable means of bird identification. With a tool that somewhat resembles a paper punch, small, numbered metal clips can be attached to the wings of individual birds. The bird numbers can then be logged into a flock book with the background of each bird going into a mating history.

The toe punch will work as a marking method for even newly hatched chicks as they are taken from the hatcher. As they are removed from the hatcher tray or shipping box each chick can be given round punches in the webbing between the chick's toes. As the chart at the end of this book shows, the individual punches change in numerical value depending upon which toes they are placed and upon which feet they are placed.

In very small lots individual birds can be denoted with different toe punches. More often the same toe punch will be given to all chicks that are hatch mates or are of the same mating. These toe punch numbers must be also logged into the flock records. See an example of toe punch numbers at the end of this book.

Many dislike this procedure due to a reluctance to handle the little guys or report having poorly made punches close back up. To counter this, use a sharp punch kept very clean (they cost about five dollars new), make the punches firmly, and then take the time to make sure that the little flap of skin is completely removed from the punch site. With small chicks this is a near bloodless and painless procedure.

Leg bands come in a variety of types and are made from a number of different materials. The American Poultry Association and the American Bantam Association supply numbered, metal bands with hatching year codes. A number of other leg band options are available and most come in sizes for breeds from the smallest to the largest. There are even temporary banding options that can be used with quite young birds and then changed out as the birds grow and mature.

Some bands are sealed in place on adult birds with a tool, some can be mashed closed with finger pressure, and some spiral bands simply twist and slide into place. These last ones are perhaps the best and least expensive option for use with growing birds. The spiral bands come in a variety of colors and a great deal of information can be encoded onto the bird via color combinations and which leg they are positioned upon. In a pinch when I have needed short-term identification for a few birds I have even used short, plastic pull-ties. These now come in a number of colors and lengths and can easily be removed with a snip from a pair of side-cutting pliers.

With birds identified and birds combined in the mating pens there are still factors to consider. It is a fairly common practice during the preceding winter (or earlier) to begin assembling breeding pens for the following year. Exhibition birds are hatched as early in the year as possible to secure best feathering for the show season. With early hatches, it can be determined earlier which matings to continue and which matings should perhaps be recombined or even terminated.

It is best not to save eggs from a breeding group until the male and female or females have been penned separately for at least two weeks. Supposedly, the hen's breeding tract will clear of earlier fertilized eggs within seven to ten days of the other males being removed. Still, I recall a few years ago when I started saving eggs from a Barred Rock mating pen ten days after the new cock was introduced. In the first hatch I had a dandy little Barred Naked Neck and sold the whole lot as crossbreds at a local swap.

In the numbers building stage and beyond you will be confronted with all sorts of variables most of which you can just mark up as another part of being in the chicken game. Last year we hatched several dozen Rosecomb Rhode Island Reds and well over ninety percent were pullets. Most would consider that a real boon, but we were actually sorely in need of a few backup males and enough cockerels to fill orders for small breeding groups. During other years, with other breeds, the hatches have been cockerel heavy. There have been years with poor hatches followed by years of good hatch rates from the same birds. We have had years in which fertile eggs weren't produced until late in the season or when hatches were intermittent.

Good matings will generally consistently repeat and some I have made totally unchanged for two, three or four years in succession. That said, in one of our experiment pens stands now a Red Barred Plymouth Rock cockerel. He was plucked from a box of Barred Rock chicks shipped out by an old line hatchery from the state of Iowa. We will mate him to White Rock females hoping to produce more cleanly marked Red Barred birds. While never common in my lifetime, I have read of both Red and Buff Barred Plymouth Rocks and must conclude that some of that blood re-emerged in this instance and for any one of a number of possible reasons.

Eggs and poultry meat as commodities are no longer the route to go for the small-scale family farm producer. Their production is controlled from incubator to retail case and the only real profit now taken on them occurs at the final point of sale — the retail counter. Small farm flocks must then be built on the demand of a market that is not being served or satisfied by the current retail trade. This market must be one that has the disposable income to fairly compensate the smaller, truly artisanal producers of distinctive eggs and poultry meat.

What now has to be discerned is just how much is poultry fad and how much has the real legs for the long haul. For example, I do think that the dark brown-shelled eggs do have a future, but only if truly productive laying strains of Welsummers and Marans are developed.

The focus of the heritage bird producer has to be on end products that will set him or her and their birds clearly apart from the commodity driven mainstream. Is the purported "pheasant-like" taste of the Dominique documentable? Is it marketable? It has been demonstrated that there are quite discernible differences in the meat from different breeds of hogs and cattle. Why hasn't similar research been done with poultry meat and eggs from different breeds?

The purchase of a box of twenty-five baby chicks and the passage of six months or so will not make you an heirloom producer. There are producers who have been working with some of these breeds for forty years and more.

Through their good efforts some of these heirloom breeds have been kept alive despite a most organized effort to dismiss and then debase them. They are nearly all farmers' fowl and have fairly earned their spurs.

This is not a place for dabblers or fad chasers and their keeping probably needs to be balanced out with other, more traditional breeds, and even other poultry species. Some of these breeds are going to need many years of work and, sadly, I suspect even a few more breeds will be lost.

Take them up only with a very clear vision of what they can and cannot do!

CHICKEN ASSESSMENT FOR IMPROVING PRODUCTIVITY

ONGOING SELECTION OF BREEDING STOCK

This outline for selecting desirable production traits in chickens was developed as part of an American Livestock Breeds Conservancy pilot project to recover breed production characteristics of endangered poultry. These guidelines are from well-established parameters developed by "old school" poultrymen, as documented in some of the early to mid-20th century poultry texts. This once commonplace knowledge and practice has become unknown to most modern chicken farmers due to the ready availability of chicks that can be purchased from large hatcheries.

The following information can be used by the producer to identify birds that would be good candidates to retain for breeding stock. Keep in mind that any bird that is selected for breeding must also meet the established historic standards for the breed. These historic standards were written at a time when chicken breeds were being used for commercial production within several production systems. Input from the top breeders of each breed was used to establish the particulars of size and other qualities that would produce the best specimen for the role each breed was designed to fulfill.

A producer needs to retain far fewer males than females for breeding stock. With this in mind, rigorous selection of the males is an important component to a sound, breeding program. It should also be remembered that adult size is controlled by the size of the female stock — undersized or otherwise poor quality females should not be retained. Therefore it is better to hatch more chicks from fewer hens than to retain undersized or poor quality hens.

Understanding the historic role of the breed being evaluated is critical to success if the breed is to serve in the purpose for which it is designed. Dual-purpose breeds, such as American breeds like the Buckeye, Delaware, New Hampshire, Plymouth Rock, and Rhode Island Red, should have equal consideration given to egg production indicators as to meat considerations in order to retain their practical usefulness. Egg-laying breeds, such as the Leghorn, Minorca, or Ancona, should have more emphasis placed upon the sections of their bodies devoted to egg production, but will still benefit from a sound overall appraisal. Because there is a link between the breed, the environmental system for which it is designed, and the products the breed is meant to produce, selection of breeding stock should favor those animals that excel within conditions in which the breed is meant to be raised. In other words, when planning to use a breed designed for range-base production, an animal that grows quickly in confinement should not be favored over a slightly slower growing or smaller animal that was grown on pasture. In such a case, the differences between the production systems, and not genetic differences, may cause the differences between individuals. Comparison of individuals within the same system does correlate, to a large extent, to selection based upon a measure of genetic makeup, and thus breeding potential and quality.

In 2006 and 2007, ALBC worked with Buckeye chickens and developed a model for the recovery of production characteristics within endangered chicken breeds. Through this work, it became clear that the key to success was in selecting birds for six basic qualities: rate of growth, mature size, egg-laying ability, breed type, color, and fertility and vigor.

BASIC QUALITIES FOR SELECTION

RATE OF GROWTH — Speed of weight gain influences profitability and can indicate strength of the immune system of the bird as well as suitability for system of production. It is a well-documented fact that both excessively fast growing and extremely slow growing poultry have less robust immune systems. Excessively fast growing birds can be more prone to diseases because of thinner gastrointestinal tracts which allow both faster nutrient uptake as well as easier penetration by bacterial, and possibly viral, agents. Historic level of productivity is a good guide for optimum rate of growth for the breed under consideration. For many of the American breeds, historic rate of growth, i.e. the time it takes to grow to processed size and weight, is between 12 and 18 weeks of age — with the majority of breeds falling toward to the higher end of this range. Rate of growth must also include fleshing; as it matters not how large and heavy a chicken is if it has no flesh when it is processed.

MATURE SIZE — The ideal weight for each breed is outlined in the American Poultry Association's *Standard of Perfection*. The figures given are an ideal with a permissible range of plus or minus one-half pound. It is important to remember that mature size also refers to the fleshing in the economically important sections of a bird's body. In order to reach ideal mature size, the bird should reach desired weight and have ample flesh in the sections important for that breed. The weights and body proportions given were determined by the top poultrymen of their day, who used these breeds commercially and identified the most productive fowl for the systems and used this knowledge to create the standard descriptions.

EGG-LAYING ABILITY — Chickens that do not lay eggs do not reproduce. Egg-laying ability is an important economic consideration — fewer breeding hens are needed to produce a given number of offspring when the hens lay large numbers of eggs. Selection for egglaying ability is a trait of primary importance in egglaying breeds, of significant importance to dual-purpose breeds, and of some importance even in meat producing breeds — though high egg production and high rate of growth are not completely compatible traits. Most breeds will begin to lay at around six months of age. It has been found that selecting for earlier production reduces adult egg size. Rhode Island Red breeders have found superior overall health and production in pullets that begin to lay at or near six months of age over those that begin to lay at eight to nine months of age.

BREED TYPE — As stated above, the American Poultry Association's *Standard of Perfection* outlines the ideal of the breed. Type is comprised of body shape and conformation and is important because it affects the size and shape of the internal organs and the distribution of flesh, and thus the breed's suitability for the system of production. Breeds like the Wyandotte, Buckeye, and New Hampshire have rather compact but deep and wide bodies. Such bodies are ideally suited to retaining heat, so it should be no surprise that these breed do well in cold regions. The Leghorn, Ancona, and especially Minorca tend to be rather longer and narrower proportionally and are well suited to hot climates, reflecting their Mediterranean origins. But Leghorns are also designed for egg production primarily, whereas Buckeyes should produce eggs and meat. So breed type is an important consideration for purpose as well as regional adaptation.

COLOR — You can't eat color. So why should any consideration be placed on this trait? Color can and does impact a breed's suitability for different systems of production. For example, while white chickens are healthy and will do well on pasture, white chickens are slightly more prone to predation. Color can also be an indicator of breed purity, and therefore an indicator of the genes that gave the breeds the abilities for which it is noted. Historically, individual strains sometimes had slight differences in color which were valued for giving the ability to identify and discriminate – a breeder might recognize that a given bird was not a pure representative of a particular strain, and therefore may not produce the desired results expected of that strain.

FERTILITY & VIGOR — No animal that exhibits a lack of vigor or good health or which proves low in fertility should be used as breeding stock. The only exception is when salvaging a rare line, variety, or breed. High levels of vigor and fertility are the foundation upon which economic value is built. Both of these traits are of the utmost importance and together they give the breed the ability to withstand challenges — including inbreeding or disease.

CULLING

An entire book could be written on culling. It is the single most beneficial practice that poultrymen can use to better the quality and health of their flocks. An old saying is that the best tool you can use to improve the quality of your birds is an axe! This applies to immune function as well as production, type, and feathers. A well-known Leghorn breeder and poultry judge, Mr. Richard Holmes, used to tell a story about a master breeder of White Leghorns who in his early years hired an older poultry judge to come and cull his flock. The old judge locked himself in the poultry house and started catching and killing Leghorns. The story goes that the discards came fast and heavy. When the judge was finished the breeder had only one trio left out of 150 birds. The breeder later commented that from that day forward he made progress!

DISEASE RESISTANCE

- The old-time breeders used to say to never use a bird in the breeding pen that had been medicated that year. While the bird may seem healthy, that the bird suffered disease is one indicator of low immune function. Also, in some cases of disease, the symptoms may have dissipated but the animal may not have completely recovered.

- Culling all birds that become sick is one way to positively select for disease resistance in breeding stock within the region in which the flock is located. Many poultry breeders have found that after a few generations of culling all sick birds, illness will no longer be found in the flock. This practice should not be expected to work for highly pathogenic diseases.

- Master breeder of Brown Leghorns, James P. Rines, Jr., said many times, "Your flock will have only what you tolerate." This saying can be broadly applied to all aspects of breeding, including disease resistance.

VIGOR

- Selecting for vigor requires selecting from amongst the dominant cockerels and pullets when choosing future breeding prospects.

- Select male and females that have bright red combs without dark tips. Dark tips can be an indicator of heart trouble.

- Select birds with bright, strong eyes with well-formed irises and correct eye color for breed. Some diseases, such as leucosis, prevent the iris from forming a nice round shape and may leave the eye off colored.

- Very active and animated individuals are often highly fertile and vigorous.

- Birds that have thick, well-fleshed shanks for their breed tend to be more vigorous.

- Fertility into old age and longevity are indicators of vigor.

THE LAW OF TEN

Quality versus quantity. It is an old breeding axiom that improvements and high quality are found in small portions of a population. The law of ten states that in order to find one good representative, ten must be produced; to find one great individual, one hundred must be produced; to find one exceptional individual, one thousand must be produced. Retaining only the top ten percent each season will allow a breeder to make progress toward their desired goal.

As ALBC began work on the Buckeyes, culling was organized such that the best representatives from each mating were retained so that no mating was favored over all others, even ones that produced more superior individuals. This approach allowed progress in productivity to be made while still retaining much needed diversity in the breeding population. The law of ten was also applied by retaining as breeding stock only those individuals that made it to the top ten percent of those produced that year. Three years of breeding represented significant progress and overall increase in quality of the stock produced.

SOME OTHER BREEDING POINTS

- Monroe Babcock, creator of the Babcock B2000 commercial egg-layer, recommended using hens for breeding that lay before 10 am. He noted that such hens tend to lay more eggs, and are generally healthier and long lived.

- Eggs from the best layers tend to hatch as well or better than those from poor layers.

- Evidence indicates that breeding from only two-year old and older hens increases longevity and reduces mortality within a strain.

- Keep track of your most productive hens. Sons from these hens should be favored during selection and mated, when possible, to hens that lay near the top of your flock's ability in order to produce highly productive offspring.

- First- and second-year egg production should guide retention. Hens with high records from these two years should be used as long as productive.

- Malposition of chick or air cell accounts for chicks that do not make it out of the eggshell – this is highly heritable. Cull all chicks that are unable to hatch unassisted.

- Overly large eggs result in chicks that have faults such as extruded yolks and other incubator-related weaknesses and hatchability problems. Placing too much emphasis on large egg size can result in poor hatchability for your flock.

- Rough, coarse comb texture can be linked to reduced fertility.

APPLYING SELECTION

Though there is an annual cycle to breeding, there is also an entry point and a desired goal when improvements are needed. Below is a sample breeding plan for dual-purpose poultry, which can be adjusted to fit the particulars of any poultry breed.

Year One

- Hatch. If attempting to make progress, it is best to hatch in sufficient numbers to allow selection and retention of superior individuals rather than maintain the status of the strain. By understanding the law of ten, it is easy to see that a target of thirty offspring should be set to simply find one good trio — sixty to find two trios to retain as breeders.

- In the first year, selection should be harder on male offspring than females when starting from a small group. In the Buckeye work, the first year produced only five pullets – thus all five had to be retained as breeders for the following spring.

- Do not weed out different lines. Try to hatch enough offspring from each line so that diversity may be conserved while selecting the top ten percent for retention.

- Evaluate the young birds for rate of growth. For the Buckeyes we choose 16-week weights and evaluations as ideal because this age was a good choice for selecting potential breeders, and because the young birds would be at or near processing age. Superior birds should be banded for retention.

- Keep records of egg laying, fertility, and molting ability of the parent stock used. Mark individuals that excel in any of these qualities and retain for use again in year two.

Year Two

- Set up matings to avoid close inbreeding and to make good use of the genetic diversity available.

- In late February appraise the hens and mark those that have begun production and which indicate potential superiority as appraised for egg laying ability. These should be retained for continued use as breeders.

- Also in February, appraise the males for the potential to pass on good capacity for egg production in their daughters and make notes as season progresses on fertility.

- Adult weights should be taken in March or April, as at this time all cockerels and pullets retained from the previous season should be approaching adult weight. Compare weights with standard requirements for the breed.

- Hatch. Again, allowing for selection of the top ten percent. Ideally, produce 10-30 chicks per female used so that hens can be evaluated as well as males for the quality of their offspring.

- Evaluate young birds at 16 weeks of age. Mark superior individuals and compare to last year's appraisals.

- During the late summer, observe the molting of the adult birds. Make note of individuals that molt in late August or September and those that drop all their feathers at once. Preference should be given to these individuals.

- Throughout the year make note of egg laying, fertility, and molting ability of parent stock.

- Cull parent stock as necessary to retain quality for egg production, rate of growth, fertility, and diversity while fitting flock size to facilities.

- Appraise all retained breeders for proper type for breed — special emphasis should be placed that females have correct type.

YEAR THREE

- Set up matings.

- Appraise cocks and hens in late February as before.

- Weigh adults.

- Plan hatching to facilitate desired number of offspring and good quantity from each hen.

- Evaluate young stock — same age as previously. Set minimum requirements for rate of growth that all young must achieve to be retained – this will likely equate to year one's better rates of growth.

- Evaluate molting of parent stock as before.

- Cull adult males based on a combination of rate of growth, capacity for egg production, adult size compared to the standard for breed, fertility, and vigor. Color may be included at this stage if numbers allow.

- Appraise all retained breeders for proper type.

YEAR FOUR

- Continue as in year three, but minimum requirements should be increased.

- Eggs may be culled for size and shape before being placed in the incubator, though pullet eggs should be compared only to other pullet eggs and not hen eggs for size.

- Consistency of size, rate of growth, and color should be more apparent in young stock.

- Color and plumage quality will certainly be considerations for young males this year.

- Parent stock that continues to meet requirements should be retained in the breeder flock as long as viable.

YEAR FIVE

- Continue as in year four.

- Eggs may be culled for color before being placed into the incubator.

- Females may now be culled based on color and plumage as well as males.

- Minimum requirements may need adjustment.

SUMMARY

Breeding is not simply a static, intellectual pursuit, but requires a certain level of creativity and flexibility. The choices made by the individual breeders not only help to mold a strain of poultry, but they can be a source of pride and satisfaction for the effort of managing the breeding stock. Breeders should feel empowered to tailor choice of selection criteria to fit their desired goals and needs.

However, there are some basic ideas that should be kept in mind as you progress. In the first year of selection there is much advantage to emphasizing rate of growth and body capacity. Males in particular must be viewed not only for their obvious positive qualities, but also for their potential to produce both excellent sons and daughters. To that end, appraising males as if the were hens in production of eggs greatly supports the maintenance of egg production within the strain. It is better to keep your second best cockerel for breeding if he is close to the best male for rate of growth and fleshing, and if he has superior width in the back and a larger distance between keel and pelvic bones. Such a male will produce offspring that will grow well and lay well. It is also best to give small consideration to fine points, such as color, in the first few years. As progress in other areas is made, emphasis can be added first to male offspring and in later years to female offspring.

There are some cautions worth considering that help to make sound long-range decisions. Intelligent breeders must keep in mind their long-range goals and avoid shortcuts so that the final result is a strain that has the diversity to stand on its own while producing as expected. Much faster progress can be made by discarding matings that do not excel for the traits focused upon, but later the diversity these "lines" lend is well worth the effort to bring them up to the levels of the other lines of the strain.

— *Don Schrider*

Originally published by the American Livestock Breeds Conservancy, PO Box 477 Pittsboro, NC 27312 USA phone (919) 542-5704 fax (919) 542-0022 albc@albc-usa.org, www. albc-usa.org.

· CHAPTER 6 ·

CARING FOR YOUR BIRDS

BARRED PLYMOUTH
• MALE •

BARRED PLYMOUTH
• FEMALE •

WHITE WYANDOTTE
• MALE •

WHITE WYANDOTTE
• FEMALE •

PERHAPS THE BEST WAY TO BEGIN THIS IS BY SAYING THAT I HAVE SEEN VERY LITTLE THAT A CHICKEN WOULDN'T EAT. Their ancestors lived upon just about anything that could be found at what might be termed "jungle fowl level."

Chickens seem to think that, from just below the surface of the earth to what they could fly up to reach, just about anything was fair game. I have lifted many a feeder or nest unit and watched hens dine sumptuously on any baby mice to be found there. From bits of grass to worms, fruit, insects and whatever they could get their beak around — it went down the hatch.

Chickens on range are actually very poor utilizers of green plant growth. They eat small bits of plant life and in winter will even benefit from access to a bit of good, leafy alfalfa hay. However, on range they draw the greatest benefit from insects, bits of spilt grain and grain found in livestock waste, and dropped fruit. These along with plentiful vitamin D from the sunshine and a few things that can be dumped into a category denoted as "containing unknown growth factors" really give outdoor grazing poultry a boost. Aside from the insects, grains and leafy plants, I have seen true omnivore chickens take out the odd lizard or small snake or two over the years.

Heirloom chicks aren't your basic chick day specials from the farm supply store, but they aren't hothouse orchids either. To reach their fullest genetic potential what they need is proper sustenance and well thought out care.

For perhaps the first two-thirds of the twentieth century more nutritional research was done with chickens than perhaps any other livestock species. The first great challenge was to keep them productive throughout the winter months.

Supplemental lighting was key to that, but nutritional support was no small matter either. Early efforts to bolster winter laying rations were broad and varied and many of these still have value to the small-scale producer today. For example, to replace browse-on greenery there were many efforts to produce and feed sprouted grain. Some of those early seed sprouters were quite elaborate, but sprouted grain was not an easy item with which to formulate rations. Other measures were basically to add extra touches or side supplementation to the feeding of grain-based rations. One longtime practice was to suspend cabbages just above the birds' heads in winter quarters. Through some slight effort birds could acquire a bit of greenery and release a bit of pent up energy.

We have often used the longtime practice of topdressing winter rations with a bit of cod liver oil. It adds protein and a number of other nutrients in a time honored, fairly simple and inexpensive way. Fill the laying house troughs with the birds' regular ration and then pour a moderate line of the oil down the center of the trough two or three times a week. Do not use the flavored varieties created for children on poultry.

Ration tweaking has been a longtime practice among small scale, purebred poultry producers. Most have a trick or two for use in each and every season of the year. I pull dandelion plants (roots and all) and give them to our birds as a bitter green throughout the spring, summer and fall. As long as the birds consume them eagerly then they are providing for a nutrient need.

A hard and fast rule to be noted here is that feedstuffs are never a place for cost cutting. They must be fresh, properly formulated, and in a form that encourages consumption and reduces waste — such as crumbles for young birds which progress to small pellets as they mature. Cut bird numbers before feed quality is reduced.

Sadly, many feed companies put their poultry nutrition work on the shelf when the big, commercial flocks begin to emerge. In many instances poultry nutrition became a poor afterthought, became badly dated in formulation, and otherwise suffered from simple neglect. When retailers have to blow the dust off the sacks to read how to feed the rations it is probably best that you look a little further down the road for a feed supplier.

Thankfully, a handful of the major feed companies have kept alive their poultry research and product development. The real test to see if a feed company is really committed to chickens is to note the variety within their poultry feed lines, the number of secondary products they offer, and how they market to and treat their poultry feed customers. The local grain elevator we buy from has seen a marked increase in poultry feed in the past years. It now even appears to exceed swine feed sales there.

Good modern poultry feeds pack a lot into those brightly colored paper bags. They can generally be bought and used with great confidence.

Easily accessible now are starter, grower, and laying rations for chickens; starter and breeder rations for gamebirds; and even special mixes for waterfowl and turkeys. The heirloom producer may actually draw from all three of these groups and then opt for even further ration supplementation measures.

A few years ago I sought a feed product to help bring up the yellow color on the feet and beaks of some of our birds. I believed it would have to be rich in carotene (actually the chlorophyll in alfalfa will help with this). I asked our feed dealer for help and he forwarded the question up the ladder in his company. Their recommendation was to use their ration for maintaining the pink color on captive flamingos. It was formulated with shrimp and crab shells.

Many of these new poultry rations are now so complete in their formulation that you no longer even need to offer grit or oyster shell when feeding their birds. Actually, now there is some real concern about whether oyster shell is even usable by chickens.

If you do need to feed a grit product, or feel the need to do so, cherry granite grit in the size appropriate to the age of the birds is the best choice. From time to time we will go to a nearby creek and scoop up a bucket or two of sand to dump into our pens to give the birds something to scratch through and gather some grit.

When consumed, grit moves to the bird's gizzard where it is compartmentalized to help with the breakdown of grains in the digestive process. Read and follow closely all company directions on how best to use grit with their feeding products.

A good heirloom poultry farmer is like a good chef. He or she seems to know just when the birds need a little extra soupçon of something. The heirloom and rare breeds often seem to need something in the way of a boosted ration to compensate for a lack of vigor or libido that can stem from an overly narrow genetic base.

Quite often it is a nutritional problem that sets the stage for a great many health and performance ills. Ration boosting can be quite simple and involve the most basic of items. The key is to use them in a timely manner and not in a way apt to cause any further upset or stress. Sudden, drastic changes in rations, even if intended for the best, could be very upsetting to a bird's well being.

A few years ago I was at a Midwestern farm show when two ladies come to me nearly in tears. They had just bought a couple of trios of very high dollar heritage breed turkeys and then had been handed a formula for a most complex turkey ration.

It ran to over twenty ingredients, some quite exotic, and nearly all were to be one hundred percent organic. No doubt, it was a good ration but would cost hundreds of dollars per ton to duplicate if dependable sources could even be found for all of those different ingredients.

Poultry rations can range from quite elaborate to bare bones simple. A most wise, early investment for any poultry venture would be a pre-World War II copy of the nutrition classic by Frank Morrison, *Feeds and Feeding*. In it can be found the nutrient content of hundreds of feedstuffs, guides to ration formulations, and numerous sample rations. This book might appear dated, but most of it is bedrock data and the feeding programs outlined were put together when the great majority of farm poultry flocks were built on larger breed birds and were being held in simpler facilities.

When it is realized that even a hen in heavy lay is only going to consume four to six ounces of feed daily – the bottom line is that bird numbers have to be quite large to justify any sort of investment in on-farm feed processing equipment. With modest numbers and birds in small pens, commercial pelleted and crumble-type rations certainly help to keep things simplified.

Today, poultry feeds are a rather specialized, although growing field. Just a few years ago for example, chick starter feed would often only be available but for a few months each year. Duck and turkey rations could not be had in many regions and other poultry feeds were special order items with rather substantial minimum purchase requirements.

Along with good variety, you should be concerned with the freshness of their feed supply. Feed left stacked for extended periods in the far corner of a warehouse will suffer nutritional and quality breakdowns. Problems can develop with dustiness (fines), caking, water damage, vermin contamination, mold, and the like. Buying in modest amounts that will be used in rapid order will do much to assure that fresh supplies of feed are on hand at all times. This is a practice that can also help to even out the highs and lows in feed prices.

All birds consume feeds in such small amounts that every measure possible should be followed to assure that they get maximum benefit from their feedstuffs. With smaller numbers the higher costs of commercially prepared feed products is generally offset by convenience and versatility in their use.

OPTIMIZING COMMERCIAL POULTRY RATIONS

The following are some tips to get optimal value from commercially available poultry rations.

• Smaller pellets, sometimes called mini-pellets, seem to be consumed better by started and adult birds than meal or crumble type rations. Pelleting has also reduced feed wastage for us as pellets tossed or scratched from the feeder are often retrieved and eaten by the birds later in the day. I would like to see a grower feed offered in a pellet and would not be surprised to see the development of a gel-type starter product one day. A gel-type hydration product is now available for birds of all ages when in transit.

• Unmedicated chick starter feeds are now much more widely available than even a few years ago. A great many will carry some level of the product Amprolin that is used for the control of coccidiosis. This product is said not to harm baby waterfowl.

• Follow feeding instructions to the letter. The same is true for all drug products. Do not make any abrupt changes in ration types or sources. Achieve such changes gradually over a period of three to five days.

• Many poultry rations are now built entirely with vegetable matter that can add to costs. Also feed with higher levels of crude protein may be harder to obtain. Some producers will even provide added vitamin and mineral supplementation when feeding with these products. In times of stress or change we always add a vitamin/electrolyte product to the birds' drinking water. Birds will drink when sometimes they won't eat.

• Many still believe a source of animal protein is needed if the birds are to achieve optimal levels of performance. Birds left on their own will eat a lot of "meat" whether it is worms, insects or any other creepy crawlers. If feeding from a totally distinct species such as fish, I feel there should be no real concern for any special crossover from potentially harmful organisms.

• Most feedstuffs for young birds are designed to be fed to birds of specific ages. Follow those recommendations and possibly even extend them a bit for late-season hatched birds.

• Commercially available feedstuffs from a company working to stay current in the field is a good base from which to begin building an heirloom breed feeding program. There are supplementation measures and specialty products that individuals can use to further ramp up performance from their birds.

Chick Nutritional Needs

Their nutritional needs begin when the chicks arrive in the mail or are removed from the hatcher and placed in the brooder. Those chicks that have had a rocky trip through the mail or are out of a slow hatch will need that bit of extra feed as noted above to make a successful launch in life.

The best place to begin giving them this boost in life may be through their drinking water. For the first couple of days add four to six tablespoons of simple white sugar to each gallon of drinking water. This gives the chicks a quick energy boost and they will often drink when they won't eat. Two to three times each week we also add a vitamin/electrolyte product to the drinkers in the brooders.

There are a number of good vitamin/electrolyte products available. Some are formulated for poultry (all will work) and a couple even contain probiotics. There are some who believe that this product can be overused. They contend that chicks may not fully utilize it and overall water quality can be adversely affected if overused. We have not had these experiences, but do change the water frequently and clean the small waterers often. They can be dipped and rinsed often in a mild chlorine bleach solution (one ounce of bleach per gallon of water).

Longtime friend Dane Hobbs from down Texas way has us doing something else that works well to keep the water fresh, too. Two or three times each week I will add a cap full of hydrogen peroxide to each gallon of drinking water. This seems to improve water quality. It freshens the water due to the extra oxygen molecule, keeps our waterers cleaner, and even seems to help check that annoying green film that forms in waterers during warm weather.

Most baby chick feeds now are quite complete feeds onto themselves. They are sold in various small particulate, "crumbles" forms. An acquaintance with gamebirds and bantams will run starter feeds through an old blender to break down particulate size even farther. Just don't beat it to dust. Baby chicks can consume only very limited amounts of feedstuffs and for very best results they should be as nutrient dense as possible. With bantams and some standard chicks that you know are going to need an extra boost you can start them on gamebird starter or a blend of gamebird and regular chick starter.

This is a hot, hot feed and you need to watch the chicks closely for any related problems such as pasty vents. It is always best to start them directly on this product if you are planning to use it. If you must make a transition to it do it ever so gradually over a period of days.

Good starter and early grower rations can be quite complex in their formulations along with being nutrient dense. This is due to the extra nutrient needs and quite limited ability of a baby chick to consume. It explains the higher cost and the convenience factor benefit of using starter feeds for one with modest poultry numbers.

Currently being offered are a number of broad purpose rations that are supposed to be used as a starter/developer or base beginning ration for a number of poultry species. They are sound in their formulation, but I believe they are perhaps trying to fill too many needs out of a single bag of feed. I have used one to finish developing some heavy breed birds late in the season when the higher protein level seemed to best suit their needs.

Our best experiences have been with finding good starter/grower ration and staying with it until the birds are at least twenty to twenty-two weeks old. No, I don't own shares in a feed company, but I do recognize the importance of keeping feeding programs as simple as possible and keeping that all-important bite-after-bite consistency needed in young bird rations. Even if raising baby chicks with broody hens these higher quality feedstuffs are needed to be offered to both. My barn banties of nearly fifty years ago did raise their small broods on little more than found feed, but both growth and survival rates suffered for it.

Early in the breeding year our challenge with some of our heritage birds is to get them laying well and producing fertile eggs as soon as possible.

Some breeds, like the Sumatra, do not truly hit their reproductive stride until warm weather and the accompanying longer days of daylight. Others need a bit of a high protein push to get them laying (and laying fertile eggs).

There are a number of ways to boost protein levels in breeding rations without breaking the old ration. First, be sure that the birds are up to making the fullest possible utilization of that extra protein. They should be free of both internal and external parasites, well housed, and in a true weight gaining condition.

Perhaps the simplest means of stoking their reproductive engines is to gradually switch them over from their regular complete feed to a complete, gamebird layer ration. It won't be cheap, but it will get the desired results. Time after time I have seen just this one change jump egg production and increase hatchability dramatically.

BOOSTING BREEDING
BIRD PROTEIN LEVELS

There are a number of other ways to boost the protein levels going into the birds in your breeding pens.

The concept of feeding animal-based protein is somewhat controversial right now, but no other protein and mineral source is more nutritionally dense and easily digested. I have to believe that there are few real health risks if the protein is of high quality and from other, very distinct species.

Some pet foods such as cat food that is rich in fish protein or the moist dog foods that resemble hamburger, or even some small particle catfish feeds can be offered to breeding birds. Offer them in small quantities two to four times each week in amounts the birds will clear up quickly. For example, form the burger type dog food into pea-sized balls and give each bird one or two per feeding.

Also possible is to use actual hamburger. This is a good way to use up hamburger with freezer burn or which has been stored too long. Again, form it into pea-sized lumps to feed.

Milk or even yogurt can be offered to the birds in a trough or small pan a time or two each week. Figure about a teaspoon of yogurt per bird or offer only the amount of milk they will clean up in an hour or so. Make your own yogurt or contact local stores to see what they do with dairy products that are near or past their expiration dates.

Most modern laying crumbles and pellets are quite good complete feeds and contain everything in them right down to a grit source. I like the pellets as there is less waste with them. They do not even need to be supplemented with scratch grains.

Many like to feed scratch grains to increase energy levels going into the birds during cold and inclement weather. With deep litter systems in loose bird houses a bit of grain is often tossed atop the floor litter to encourage the birds to scratch in it and turn it for better decomposition. If grain is being offered as an energy booster in cold weather, offer it later in the day.

The birds actually seem to favor the grains over their layer rations. Unless it is offered in a very controlled manner the birds will overconsume the grain, eat less layer rations, and the resulting egg production will suffer.

The old timers had a trick or two up their sleeves for boosting winter egg production that will work in other seasons of the year, too. It was their common practice to topdress the hens' feed with a healthy drizzle of wheat germ or cod liver oil. This measure adds a great many nutrients to the birds' ration.

A variation of this practice is to take a gallon of oats and mix into it a quarter cup of wheat germ oil. Each breeder would then get a teaspoon of this two or three times a week. It may take some time for your birds to readily accept oats. The following week mix a quarter cup of cod liver oil into the same amount of oats and feed as above. Do not use the flavored varieties of cod liver oil.

To either of the above blends, a cup of yogurt can be added to further increase its nutrient content. This is also good feedstuff for birds that have been on a treatment course with an antibiotic product. It can help to restore valuable bacteria to the bird's system. Bacteria are needed in the gut for the digestive process in birds as well as humans.

Broilers on range really get very little benefit from green feeds. The real pluses for them are really the fresh air, sunshine, and exercise. These fast growing birds need access to a carefully formulated ration that accounts for their growth rate and frame development. With the "fast broilers" many producers now limit their birds access to feed after they are about four weeks of age. They will pull the feeders away from the birds from roughly seven in the evening until seven the next morning.

Shop carefully for broiler growing rations and match the rations carefully to the breeding behind your birds. Don't assume that because they're on grass that they will self-balance their rations. What goes into that feeder has to meet one hundred percent of a fast growing bird's needs.

Many don't think of it as such, but drinking water is also a feedstuff. In fact, it is one of the most important of all feedstuffs.

Water should be before the birds constantly and in containers that will keep it clean, fresh, and appealing. In very hot weather it may even be necessary to offer the birds fresh water two or even three times a day. In winter many will offer warmed water twice a day. In cold weather I'm a firm believer in the old practices of giving everything a good drink and plenty to eat before the long winter night falls.

Also, just as with their rations, some producers like to tinker a bit with the drinking water. I have already outlined our reliance upon vitamin/electrolyte products in the drinking water for all classes of birds. We use this product almost daily in one pen or another and reach for it at the first sign of any sort of stress or health ill. Just as with baby chicks, we will use hydrogen peroxide in the drinking water of both growing and breeding birds.

I have also encountered drinking water infused with a number of different things. As noted earlier, white sugar will boost energy levels and can be used with debilitated birds as well as baby chicks. Two other common infusion products are red cider vinegar and dried hot pepper.

We have used both and generally have on hand a gallon of vinegar steeping with the addition of a couple of heads of garlic and a large, dried pepper. We add two to four ounces of this per gallon of drinking water a couple of times a week. Six ounces of red cider vinegar per gallon of drinking water is held to be a good natural treatment for coccidiosis. The garlic, pepper, and vinegar mix seems to contain a natural antibiotic that invigorates our birds. We use the mixture to help them maintain condition when under stress.

Poultry are fed to improve their performance, to advance the birds, and not to simply save money. Good feedstuffs are an investment in the current and future productivity of the flock or flocks.

There is a long held observation that a sick chicken is a dead chicken, one just looking for a place to lie down and die. As a management concept it does have an element of merit.

I don't think that the heirloom producer can accept the view of the commercial sector that the birds are largely throwaway inputs – use them up and toss them aside. Still, there is a great deal to be said for that old "survival of the fittest" chestnut in the development of a chicken flock. Producers certainly don't want health care measures to become the crutch that enables ill and poor performing birds to hold on and continue reproducing. A little, well-planned and supervised "natural selection" can be a good thing.

It has been my experience that keeping the birds hydrated, well fed, dry, and comfortable is the big half of chicken health care. Still, I don't like the idea of just accepting bird losses by saying "it is just a chicken." A three to five percent death loss at most of the mileposts in a chicken flock's life has been an accepted practice for a long time.

The first key to good health care is to simply be a good observer of the birds. To that end and because I don't bend down like I once did, all of our breeding and growing pens are set at waist to shoulder height. A few of the roosters think they're just as big as me because they can look me in the eye, but we generally work that out man to bird fairly quickly.

Begin looking for health and structural problems from the moment of hatching onward. Crossed beaks, crooked keels, bent toes, spraddled legs, navel ills and the like can be discerned early on. It is best to humanely do away with these youngsters as they are encountered. Burying, incineration or composting can dispose of them and all other dead. Check into your state and local laws as they pertain to the disposal of dead livestock.

With older birds there are a number of visual indications as to a bird's health and well-being. Watch for a dim, sunken or closed eye, swollen face, a tucked stance with drawn head, pale or miscolored comb and wattles, drooping wings, or stained feathers. Their actions may also be a warning sign too. Look for birds that are standing apart from the group, not eating, or have labored or rattly breathing to name just a few.

When an ill appearing bird is detected, remove it to a sheltered pen or cage well away from other birds. There it should be shielded from drafts and direct sunlight. Tend it last each time that chores are done to prevent carrying any possible harmful organisms to other birds.

It is not possible to become an avian veterinarian overnight — few of them exist even now. But there are still a few medicinal skills and practices that can be picked up that may help.

SIMPLE POULTRY FIRST AID

Consider having a few of the more common first aid and medical treatments for your poultry. The following are from our playbook.

For scratches, cuts, tears and fighting injuries it is hard to beat a simple cleaning of the area and regular applications of Neosporin, Triple Antibiotic or other common wound ointment. Deep wounds can be a special problem, so clean them with hydrogen peroxide and then use a good ointment. These ointments seem to promote healing from the inside out. For organic producers, check whether products are certified organic before using.

Respiratory ills may appear after sudden cold snaps, in very damp weather, following transport, and may even be triggered by heat stress. They can be accompanied by nasal discharge, swelling of the face and around the eyes, gaping of the mouth, and gurgling breath sounds. In treating these symptoms an inexpensive product called VetRX has done a good job for us. We open the bird's beak and send a few drops down its throat, apply a coat of it around the beak and eyes, and then drizzle a few drops to form a reinforcing film atop the bird's drinking water.

There should be results seen in one to three days. We also apply Vicks or Mentholatum to the bird's face and around the eyes being careful not to clog the nostril openings.

Quickly launch support therapy with sugar and vitamin/electrolyte in the drinking water. Use cage covers to keep chilling drafts off of the birds. Old feed sacks can be stapled to cage fronts for this purpose and then be pulled down and burned for sanitary purposes when no longer needed.

There are a number of oral antibiotic products that can be mixed with drinking water. Their drawback is that most are created for large group treatment and must be used quickly after opening. Placing them in a sealable plastic bag and storing them in a refrigerator may prolong their shelf life a bit.

Do not change rations on an ailing bird.

Most have done an off-label use of a health product or even tried a human health product on a sick bird at one time or another. This cannot be recommended and all health product labels should be followed scrupulously. Especially note and follow all withdrawal recommendations. Some health products must not be used with birds producing table eggs.

Thoroughly clean and disinfect the isolation pen with a strong chlorine bleach solution between uses.

Sometimes it is time to admit that a bird is beyond help and has to be destroyed. Do so quickly and humanely and dispose of the body properly.

There is much discussion now about the concept of "bio-security" and how its use or lack thereof affects the health and wholesomeness of poultry and poultry products. A ranging bird is in a more open environment, but not necessarily in an exposed position to whatever the winds may send its way.

A chicken on range is not simply a bird running willy-nilly about the countryside. Birds in lots and sun porches are being held even more securely. Migratory birds possibly carrying potential health problems generally stop only for large bodies of water and feed sources such as grain fields or vast lawns.

There just aren't a lot of migratory waterfowl dropping in on small flocks of contained chickens. While multiple species of poultry may be kept on a farm, they seldom are seen penned together or even in close proximity. There is strong evidence that a spacing of just eight feet will eliminate the aerosol spread of most harmful organisms.

Further, the spread of harmful organisms such as the flu from bird to animal is accomplished in only one of two ways and neither is seen on

FROM THE FARM TO THE TABLE

Henry Noll took over Noll's Poultry Farm in Kleinfeltersville, Pennsylvania in the 1970s and has been raising quality chicken for the table ever since. From the first days of business in 1933, the Noll's family has prided themselves in providing quality meat and eggs that rival the hybrids and genetically faulty breeds of mainstream producers.

When asked what the differences were between big industry chicken and the product that one can expect from Noll's Poultry Farm, Henry had a lot to say. Henry said that the birds that are bred to develop for slaughtering within five or six weeks provide "soft" meat, whereas Henry's chickens have more time to grow and create more "firm" and flavorful products. Also, Henry commented that birds that are bred to gain weight at such a rapid rate typically have health issues like broken legs, *Mycoplasma gallisepticum* (MG), Avian Influenza, and heart attacks, while his more natural heritage breed chickens typically live fuller, healthier lives.

On Noll's Poultry Farm you can expect to see White Plymouth Rocks and Barred Plymouth Rocks, two heritage breeds that are listed as "Recovering" according to the ALBC. Henry's breeding process allows him to select the best third of the breeders for reproduction without altering the genetic makeup of that breed. Henry says that when his chicks are born they're up and moving as opposed to the unhealthy chicks of other over-modified broiler breeds that have sluggish, lethargic tendencies. And it doesn't take a genius to figure out what Henry has – the heritage breeds that have been around for over a hundred years simply have better, healthier genes than the modern crossbreeds.

While he's not one to boast, Henry did comment on a few of his prized successes. What started as a small hatchery in the 1930s has now grown into a booming business with a daily hatch average of 125,000 chicks. Henry also said that his only form of advertising is in the occasional company pen that he gives out to customers every now and again. While you're likely to hear of Noll's Poultry Farm by word-of-mouth, you won't be seeing any billboards with Henry's face on them and the only information you'll find online will be from a third party. Apparently his tasty chickens advertise themselves!

With today's get-green, go-local state of mind, it's no surprise that people are interested in getting back to basics. Henry says he's happy to see people becoming more and more interested in pasture-raised poultry and heritage breeds of poultry that haven't been so customized for growth that they can't even stand on their own two feet. While Henry admits that raising chickens with this more natural production model is no small feat and requires hard work and patience, he says it's all worth it for the quality product that he's able to produce for his loyal customers.

Henry Noll is the owner and operator of Noll's Poultry Farm in Kleinfeltersville, Pennsylvania. For more information call 717-949-3560.

American farms. The first is where there is immediate contact between poultry and swine. The hogs represent a population where organisms can more easily mutate to forms of greater potential risk to humans.

The second is where humans have immediate and prolonged contact with infected birds. In 2004, two poultry farms and four live bird markets were found in the United States to have the low pathogenic form of Avian Influenza. That same year, one high pathogenic Avian Influenza outbreak was found in the United States, the first within twenty years. In some cultures, birds and humans share workday environments for hours on end, open water sources, and even sleeping quarters. Also present has been the handling of diseased or dead birds without proper sanitary measures.

How a producer in a poultry confinement facility — a chicken gulag — can claim the high ground on this one is certainly beyond me. Large numbers, often hundreds of thousands, held tightly together in a small area are already under a tremendous stress load. They are held close to their own wastes and are totally dependent upon automated systems to keep their sea of wastes under control, their air breathable, and their environment livable.

In a confinement building, generally a great many of them are packed quite closely together, and can appear fort-like, but how bio-secure are they really? Tons of wastes are held nearby, ventilation is outside air pulled in through large fans, the birds are packed tightly on unrelenting surfaces, the buildings can be breached by vermin and small birds, and they are a most inviting target for eco-animal rights, and other terrorists.

There are a great many accounts of system failures that have resulted in huge numbers of bird deaths and even some human ones. Human abuse and neglect have also been well documented in the confinement sector. Hurricane Katrina ravaged over four hundred confinement buildings dumping thousands of birds with minimal health resistance into the environment and directly into a major, migratory flight path.

Actually, long confined birds and breeding strains are terribly lacking in vigor and any sort of natural immunity. Well known in poultry circles is the story about a flock of small breed turkeys still held in the northern United States. They represent a most desirable and valuable genetic resource, but one totally inaccessible. Simply put, after long generations in confinement our present day large-scale production turkeys have lost the ability to function and thrive outside of that totally artificial environment.

Poultry on range in the United States function in a zone that would be shared with a relatively few other avian species and most of them would be genetically quite dissimilar. Open pasture is a rather uninviting place for most species. On range they do not water from ponds or streams and they should all be indoors on the roost come nightfall. Generally now they are at least four to six weeks of age before going on range and with a fairly well developed physiology.

Birds in lots or deck-type porches are exposed somewhat to the elements, but they have little, if any, real contact with natural systems. Some of these systems are even said to be internal parasite-free as the birds are held well above ground level. With full pen enclosures there is virtually no direct contact with wildfowl possible. I have actually had hawks alight some distance from my bird pens in apparent frustration at glimpsing and hearing the birds there, but not being able to come any closer to them.

Small scale and range production is not poor practice. It may be carried out with simple facilities and modest expenditures, but it does a quite good job of bird care and is what has kept alive so many breeds for so long. Bigger and better are in no way synonymous when it comes to poultry care and production.

The large units literally have small mountains of dead birds with which to contend. Just last year workers at one of the larger layer units in Missouri were filmed loading living and injured birds into a truck with deads and layer house wastes. As to the vaunted "bio-security" of these units I am reminded of the Missouri State Veterinarian who asked if I had ever seen a truly sparrow and mouse-proof steel building. There ain't no such animal. Quite often now these facilities are staffed with some employees that may not know or follow state or federal agricultural guidelines and may pose a bio-security risk.

The heritage bird producer will benefit from a management plan for preventive health care rather than a plan of bio-security with its reliance upon fairly easily breached walls and barricades. The first step to such a plan is summed up quite eloquently in something Dad said to me many long years ago, "Don't buy somebody else's problems."

Buy only from sources well known to you and who are operating with facilities as similar to your own as possible. New birds should enter the flock only after a period of isolation and observation of at least thirty days in duration. Limit human and animal traffic around the birds as much as possible. Cover pen tops with bird-proof mesh, don't pasture them adjacent to water frequented by waterfowl, and maintain a regular program of vermin control. When taking birds to our local farmers' market we make it a practice to either sell out or isolate any that are returned home just as if they were coming in from an outside source.

What gives us the greatest comfort and feeling of security is knowing that our birds have access to fresh air and sunshine, that they are given room to maneuver freely and to fully function, and that they are fed a quality diet. I can attest that there is far less stress to life in these simpler facilities, causing the birds to develop an increased level of natural immunity. They are then selectively bred for natural thrift and vigor.

Parasite control does much to keep birds both comfortable and in a thrifty, productive condition. The birds can be affected by a number of different internal and external parasites. Fortunately, the options for their treatment and control have broadened dramatically in recent years. Still, the first step to good parasite control begins with basic sanitation.

As noted above, birds held in elevated pens, with no direct ground contact, will remain largely free of a number of internal parasites. This was one of the primary reasons for the early use of "sun porches" for growing out young turkeys for slaughter.

Early treatments for internal parasites involved putting some very potent household and plant products in drinking water or adding these products to rations. For example, pepper and garlic do have some curative properties, but are not good anthelmintics (or de-worming products). The first widely used wormer product was Piperazine added to the drinking water. It is still used to control round worms. Diatomaceous earth (DE) has been fed as a wormer, but there is very little beyond anecdotal reports of success with using it. It is a contact killer and does its work by scouring soft tissue. Dane Hobbs of Brenham, Texas has developed a number of drinking water additives for additional poultry nutrition and support therapy. His organic Immuno-Boost product has some DE and I like to use it ahead of worming with other products. It seems to increase the knock down rate and overall effectiveness. The oral Ivomec product for cattle is now being used fairly widely with chickens. However it must be the oral and not the pour-on form of the product because the pour-on is not soluble in water. To use, add $1^1/_2$ cc's of the product per gallon of drinking water and offer as the only source of drinking water to the birds for a twenty-four hour period.

Some do use the pour-on form of this product in a different manner, placing one or two drops of it to the back of the adult bird's head or to another spot on the body, inaccessible to the bird. These products will control all parasites that feed upon bodily fluids. Do consult with your veterinarian before using any of these newer generation wormer products and administer them only to otherwise healthy birds. Consult your organic certifier to get a list of certified products before beginning treatment.

There are other worming products that can be used with poultry, but you must follow all label directions carefully. Any off-label use of a product should be done only under the direction of a veterinarian. It is also best to regularly rotate worming products to prevent the chance of any product immunity developing.

Early in the breeding year our vet recommends that we treat the breeding groups with a course of the product Corid (amprolium) in their drinking water. He sells us a few ounces as a prescription product and we administer it through the drinking water over a period of days. It is a coccidiostat, but seems to have some sort of residual effect on other parasites and the birds emerge in a more vigorous state. Check with your organic regulators about the use of coccidiostats before administering.

Chickens can have a number of external parasites including several types of mites. There are a number of control products available and they too should be rotated often in their use. The long time practice for mite prevention was simply to provide the birds with an area or box of wood ashes in which to dust themselves. Do not use the ash from treated lumber for this.

In my youth a whitewash made with lime would often be applied to roosts and interior walls of poultry buildings that provided some residual parasite control. With a heavy mite infestation we will dust the birds with that old garden standby, Sevin dust. It is an off-label use, but Adams Flea Spray has often been used as a spot treatment for mites, too. Any off-label use of a product should be done only under the direction of a veterinarian. Another spot treatment is a cloth dampened with Dawn dish soap, squeezed dry, and then applied to mite-infested areas with a dabbing action.

Each producer must work out his or her own approach to health matters, but just be sure to move quickly when health problems are encountered. By selecting continuously for vigor and growth, producers are building the flock on a tenet of good health, soundness, and durability. The flock can made healthier with wise application of the breeders' arts. I don't believe in feeding a different antibiotic to the flock every time the moon changes, but there should be a number of products and practices in any poultry producer's health care arsenal. I am not purely organic, nor do I believe that we always need to use the "bigger hammer" theory of health care product selection. I favor products that work quickly, require minimal handling of the birds, do not subject them to greater stress, and are cost effective. A two hundred dollar vet bill to save even a one hundred dollar chicken is hard to justify.

In no way can all of the ins-and-outs of detailed poultry health care be outlined here, but the following are some measures and practices that have worked for us.

Basic Poultry Health Care Tips

• Have a pen or cage at least several hundred feet away from other birds in which to quarantine ill appearing or injured birds. Care for them last each and every time you tend your birds.

• Remove any birds from the group at the earliest sign of problems. Others in the group may turn on them.

• Applying supplemental heat will do much to make birds feel better. I've seen chilled chicks bounce back from a near flat state when their body heat is restored. Heat will help birds that aren't eating well to maintain some condition.

• Give birds full attention while working around them. Listen for sounds arising from respiratory ills, watch for unusual behavior like gaping, check for blood or injury, look for fecal irregularities, and be alert for anything that just doesn't seem right.

• Oral antibiotics and their use and storage were discussed earlier. There are also injectable products that can be used with chickens and homeopathic remedies. Yes, you can give a chicken an injection, generally in the breast muscle. Follow all instructions fully, store properly, fill syringes through a separate needle than the one used to give injections, and note that products in dark colored bottles are light sensitive.

• Don't wear chore clothes and footwear off the farm or around other birds and livestock or to areas where other producers may be encountered.

A friend, David Andrews, a Braggs Mountain breeding flock manager, and I have had many long talks about basic poultry health matters. It is our consensus that flock health is sometimes built by the subtraction process. Death loss can have a plus side. It is how Mother Nature manages her herds and flocks for improved vigor and the will to thrive. The birds that survive stressful times in the best condition, that grow well despite being a part of a late-hatch, that reach maturity with the least amount of tweaking are the birds from which producers should be breeding. It is nature's way of removing the poor performers from the flock before they can draw down the overall health and well being of the group.

Upon occasion some birds appear to come undone for no real reason. They go light, fall behind the group or just fail to thrive. Some even seem to defy medical and health treatments. Those given a treatment with antibiotics may need to rebuild the good flora in the gut with a probiotic product. Still, it must be accepted that some losses will occur and in the long run, some of those may be for the best in the endeavor to improve hardiness and durability in an heirloom breed flock.

There are no magic bullet health cures and part of your task as an heirloom breeder will be to become a discerning user of the products that come available. Do not invest heavily in anything until proven that it will really work in your pens. Run your own trials and do so only on very small numbers at first.

It may indeed sound funny to say, but good poultry flocks are most certainly built by subtraction. Once past the early numbers building stage with heirloom birds, when producers have to tolerate a certain number of small defects, they need to get deadly serious about quality and performance improvement.

Crack open that old *Standard of Perfection* manual and study those conformation, color, and weight descriptions until the knowledge is mastered of what constitutes a really good bird in your breed or breeds of choice. For those just beginning now, the smaller things do start to count. As a small farm producer equal weight must be given to the performance traits.

A good bird will feel solid in the hand and should appear alert without being overly flighty. I was once handed a pair of a certain breed with the proviso that I must not let them go until I was inside something with four solid walls and a roof. Selection for temperament is something that is always very much in order and if simply walking by a pen sets the birds inside into a panic then those are chickens that you can do without. For poultry to be among the smallest of livestock species I am always surprised at how many are concerned about the aggressiveness in them. Roosters have always had a reputation for scrappiness and most country children have a tale or two to tell about being "flogged" by an overly protective, broody hen.

With birds from very small, closely bred populations, aggressiveness is sometimes something for which you have to select. With excessive inbreeding and/or advancing age, libido is often one of the first things to suffer. It is nearly maddening to have a breeding pen where they just stand around, including a rooster that behaves like one of the girls.

I want to see a heritage rooster dropping that wing and turning smartly on his feet. For a time we worked with Buff Wyandottes from a line noted for spotty breeding. The original birds were sold to us with that caution. To counter that I deliberately chose breeding males with a bit of extra punch and I have a couple of small scars to show for it. Roosters that fill eggs sometimes have to fill those humans around them with a bit of caution, also.

Nearly every aspect of the heirloom flock's basic care must be predicated on their initial rarity and possible inbred status. The latter can manifest itself as slow growth, low libido, poor and slow hatchability, weak chicks, and even some deformities such as crossed beaks and bent toes.

Not every bird with curled toes is inbred (some may have just had trouble hatching). It is a strong indicator though and there will come a time when you will cull virtually every such bird for that fault alone. Not long ago I saw a small hatchery owner study a first-rate pair of Cochins for the better part of a day and then pass on them because of one bent toe on the male. Such is the level of focus that producers must bring to all of their breeders and potential breeders. A poultry yard with a lot of yard chairs and upturned five gallon buckets may not so much indicate a lazy producer as one who really sits down and takes the time to study his or her birds.

It is one thing to look at a chicken, but to really see it takes time and real focus. At many shows and markets I have attended, all I could really afford to do was look. This wasn't a lost opportunity at all as much market comparison was done. Producers need to know their birds as individuals and to see their faults as well as their strengths. As my birds grow I will work through a set of young birds at regular, two to four week intervals. In the early stages of flock building with very rare birds, producers can afford to cull only those birds with the more glaring of faults. As numbers grow, however, you need to become ever more strict and exacting in the culling process.

If flock building is going well each new generation should be at least as good and hopefully better than the one that produced it. The right pair, if given the time, support, and producer expertise, can turn out dozens of copies of themselves and even some that are better. There are some in the poultry world that have scrambled and eaten eggs that others would have paid very well for because they have already produced enough birds in the year for their needs.

There are others who will produce some birds of quality, but because of variability in the flock, environment, and possibly even their own self-doubt have had to do it by producing large numbers. One of the things that has so badly hurt the keeping of purebred poultry in my lifetime is the finger pointing and arguing that one way of doing things is better than another. Some producers with certain breeds have become very protective of their turf with some show breeders selling nothing, lest they have to show against their own breeding (someone else beating them with their own breeding would really be quite a compliment). Few seem to know what to make of the emerging groups that are using pure heirloom breeds to establish anew, small farm flocks for eggs and meat. What has emerged in recent years is not only just how few flocks of many breeds and varieties exist, but just how small many of these flocks are and how many chick suppliers dip into them. It is now widely believed that just three hatcheries now hold near complete sway over the off-season trade in baby chicks and they then drop ship for a great many other hatcheries the year around.

For some of the old varieties there will be only a quite limited yearly demand beyond those birds needed to replenish or expand the home flock. This lack of demand or popularity for some breeds and varieties has to be expected and accepted. For example, many in our area report limited demand for farm fowl that are all white or all black in color.

However, some show people seem to favor these colors. Yes, washing a white chicken is no fun, but they are simpler to breed than the many laced, spangled and penciled varieties. Personally, I've always liked the white varieties and think the white varieties of Leghorn, Rock, and Wyandotte form a genetic block of considerable potential economic importance to smaller, family farmers.

It may be a year or two in building to get to the first real good breeding trio with some of the rare varieties or even longer to having multiple small breeding pens with which to work. Some breeds are just going to take longer to breed up than others. If breeders are working with a very ornate and complexly bred bird they might also wish to consider working with a second, less complexly formulated breed as well. For someone just wanting a few hens to lay or a few fryers to eat, the heritage breeds may be impractical or may be even a bit wasteful.

We recently took up the Buff Orpington breed somewhat reluctantly, but the demand for them is certainly strong in our marketing area. A few years ago a minister in a nearby small town set about to assemble a small flock of Buff Orpingtons as he remembered them from many years before. He met with much disappointment as one line he tried laid poorly, another laid quite smallish eggs, and another ate like horses with little or no production at all. Others have reported encountering small hens, poorly colored birds, slow growers, and other performance or quality problems. We once had a trio with much promise, but the best hen consistently produced double-yolked eggs that would not hatch.

Good purebreds are neither overly complex nor overly fussy in their care. They were created to breed true and be of a consistent, proven good type.

One of the driving tenets for keeping purebred fowl is that they are in keeping with the goals of modern, sustainable agriculture. Once it has become established, a purebred heirloom flock can, conceivably, breed on for generations — both the birds and humans. It can be renewed from within.

A poultry flock can be viewed as something to be burned through quickly and replaced every few months from outside sources or something that will sustain itself and the owner's family in many ways. It must always be recognized that many of the heirloom poultry breeds were bred to be rather task specific or to perform within some rather specific environmental or economic parameters. A very popular breed of the moment, the Welsummer, is a Dutch breed bred to produce a distinctive egg and to be a rather active bird of the farmyard. Many were bred for modest production under harsh circumstances and climate, and to require limited funds for their care. The goal in their development was not to get all that was possible from a bird, but all that was possible from a bird in those particular circumstances.

Sadly, poultry producers lack many of the performance assessing and testing methods that exist with many other livestock species. The individual poultry producer will now have to compile and distill as much of his or her own performance data as possible. It will be the most meaningful of any such data as it will be true to the home flock and its production environment.

Producers will need to invest in some fairly simple equipment such as a good scale unit, egg grader, and record keeping system that will help track individual bird performance. Without a trap nesting system, true, pedigreed matings are only possible in a pairs mating system. Single hens can be penned separately and a breeding male brought to her pen for one to two hours every two to three days to keep producing fertile eggs. With this system a good male can see service with several different females.

With a system that enables producers to document egg production per female you can then select eggs for hatching from only the very best performing hens. Coupled with selection for good type, producers are then on the way to building a better and ever more productive flock. As you move up the ladder of improving the performance traits expect things to develop more slowly than in the early stages of a breeding up program. The breeding males should all only come from your best laying hens.

Egg size, is another factor often found wanting in some heirloom birds. To most people, a country fresh egg is a big brown egg. Next to small numbers being laid, small size can be a most galling problem to face. The USDA issues weight standards for a dozen eggs in the various grades and it is reasonable to expect all breeds to be laying a fairly good-sized egg by at least their second month of lay.

However this, like a great many other traits, can be carried to too great an extreme. A few years ago a number of Delaware producers began to selectively breed for the production of ever-larger eggs from their birds. All too soon problems with fertility began to emerge and a good many pullets were lost to "blow outs" as a result of producing eggs that were simply too large for their frame.

Those spending any length of time with chickens will encounter a number of egg irregularities. We have, for example, a Rosecomb Rhode Island Red hen that lays a rather longish egg with a bit of a pointed end. Still they fill and hatch well. There is an old belief that the more pointed eggs produce cockerel chicks and the more rounded ones produce pullets. What Cornish producers make of their birds that produced quite rounded eggs I cannot say.

Misshapen, poorly formed, and rough surfaced eggs are really not good candidates for the incubator. However, if these are the eggs that you are getting then these are the eggs you will have to use. For very round eggs, float them briefly in warm water to determine in which end the air cell lies. It will float with the cell end up. Knowing this they can be placed in the incubator accordingly.

By early in the twentieth century individual breeders using the accepted tenets of old fashioned selective breeding were producing hens of some breeds that were laying well over two hundred eggs per hen per year.

BOOSTING EGG PRODUCTION

Getting high egg laying numbers from hens was accomplished early in the twentieth century through three fairly basic and simple steps.

Candidates for the breeding flock were first selected for size, vigor, and overall hardiness. The tough kind and the "good doing" kind are generally always the same.

Individual bird production was tracked as closely as possible. Later breeding stock was selected only from the offspring of these best performers.

The entire flock was kept carefully culled to remove small birds, birds of inferior breed and body type, birds that did not lay over an extended period of time (a flock might be worked several times in a laying season), those that were unsound, and those in poor health.

The route to better yielding and faster growing meat birds is also selective breeding. The standard for the entire poultry meat industry now is a truly artificial creation. The Cornish X broiler as we know it now is extremely complex in its genetic formulation. It is no Fl, F2 or even F3 cross, but rather an aligning of very specifically developed male and female lines. It must be fed in a most exacting manner, would be hard pressed to survive outside of a closely controlled environment, and is most unnatural in its conformation. Its misshapen nature appears to greatly shorten its lifespan. Many believe that its monstrous breast structure can no longer be supported by the bird's nerve and circulatory systems. The bird endures only with the help of a number of manmade props or crutches. It is pure artifice and not a product of the learned stockman's hand and eye. Heirloom pure breeds that were once looked at to regularly produce table fowl in volume include Black and White Giants, Delawares, New Hampshire Reds, White Plymouth Rocks, some Wyandottes, and before them, the Javas. All of these pure breeds and their simpler crosses were sent to the showers well over a half-century ago and their heritage and potential has not been meaningfully tapped since then.

Growth rate and meat yield ceased to be meaningful goals for the few remaining keepers of these breeds. Surplus males may have found their way to the soup pot, but they were not given the support and the opportunity to compete with the Cornish crosses on any level.

Today the commercial retail market is more for "pieces" of chicken than it is for whole broilers. In my youth they were smaller and called "fryers." Now much chicken meat is restructured into strips, nuggets, and drumettes. I'm not even sure from where the idea for "popcorn" chicken came. The modern broiler is now bred to be no more than the raw ore for the chicken parts factory.

From the heritage breeds can come a meaningful, efficient to produce, and profitable to raise young meat bird. Such birds won't, thank goodness, ever fit the Cornish X profile, and will, in point of fact, vary somewhat in appearance from breed to breed. However, that distinctiveness may be a real plus for them.

Right now a lot of "black" broiler chicks are being marketed and sold to specific ethnic based markets. Some are surplus Black Giant or Australorp cockerels, some are their crosses, and others may be drawn from a variety of other black breed varieties. These are large framed, generally rather slow growing birds and cannot be considered very efficient for the task. A fair number of black birds were even produced from our Blue Wyandotte breeding group.

What heritage producers have to be about is not just tweaking the broiler concept, but rebuilding the whole bird. It is now widely recognized by informed and concerned consumers that the range broiler isn't the confinement bird. Both camps of producers are still trying to largely work with the same set of genes. It probably never should have been tried in the first place and its days on range are now clearly numbered.

The farm bird and the factory bird will ultimately emerge as two very different types of fowl. This is not to say that the farm bird will be without important economic and performance aspects. The task is to reconcile legitimate economic needs with the farm friendly and range practical nature of the needed bird and the genetic realities of heirloom poultry genetics.

The range bird market is comfortable with a three to three and a half pound dressed bird. It wants that bird to be tender and succulent but not mushy; naturally flavorful, humanely produced, and coming from a real farm. The market matters very little, if at all, whether it took six weeks or eleven weeks or even fourteen weeks for the bird to reach that weight. Their belief is in the natural superiority of a traditionally bred and raised bird.

We live in an age of smaller families for whom large primal cuts of meat and big table fowl are almost a liability. The most recent definition of "eternity" that I have heard involves two people and a turkey. Consequently big birds and big serving sized pieces are really non-issues in this trade. I suspect the real marketing challenge now would be to actually move some of those whole, elephantine broilers through supermarket showcases.

The purebred meat bird has been a prisoner in time for many decades and everything from body type to ration formulation must be reassessed and made subject to change. I think that having a harvestable purebred broiler before five weeks of age may be too much to hope for, but where would they be now if purebred broilers had received the same amount of attention as the Cornish X bird?

The eleven-week-old, harvestable purebred bird should not be that much of a reach. The eight to nine week old broiler, a point where Cornish X birds were not that long ago, is doable. I can't give the time frame for when any of this will be achieved, but the need is right now for independent producers to be selecting their largest, fastest growing, and best-formed Rocks, Wyandottes, Delawares, New Hampshire Reds, and others to start breeding toward such goals.

Such young birds need to be going on to scales to chart their growth rates and to begin guiding breeding bird selections. Poultry groups need to get behind realistic growth and yield standards for range birds and to create a clear and accepted definition of just what is a "range bird." Flavor, shear and nutrient content testing among the heirloom breeds is long overdue.

In farm circles there is an old, old adage that "the scales don't lie." Only by putting the growing birds on the scales at regular intervals can producers determine exactly how their birds and rations are performing. Advancing growth rate will take time and it is quite possible that within some flocks the growth rate will plateau before the desired rate of growth is attained.

If this does occur it may become necessary to look outside the flock for an outcross infusion of new genetics. This procedure was outlined in the earlier discussion of breeding practices. Just don't assume that you can accomplish what you want to attain by buying the first super big specimen of the chosen breed that you encounter. Very tall and leggy birds may actually work against the goal of building an improved meat type.

You may need birds with substantial frame (skeletal dimension), but who are also blocky and built a bit closer to the ground.

I field many questions about using purebred Cornish birds to develop an Fl crossbred bird that would be valuable as a broiler. This is simpler said than done. The first obstacle is the relatively small number of standard sized White Cornish to be found in this country right now. I have actually crossed a Dark Cornish Standard male on White Rock hens of fairly good size and type. These are two very distinct breeds and their first offspring were really all over the place type-wise. Most were white although a few emerged with a not too bad Dark Cornish color pattern. Some looked like White Rocks, some had the bare breast patch of the Cornish, and they all had a bit of boosted size and growth to be expected from heterosis or hybrid vigor. There was just too much randomness in the offspring of this particular first generation cross.

Others, I think, should give this cross at least a small trial if they can access good White Cornish males. The birds don't have to be show winners, but they do have to be most solid in their Cornish basics. The quest will be to find Cornish and White Rock breeding stock that truly nicks or comes together in a consistently desirable way. What may be most important in this first cross would be to produce an Fl male that could be bred back to purebred White Rock females and begin creating a solidly balanced breeding line.

My thought is that the more doable pursuit for a small-scale producer is to resolve to produce the best performing birds possible from a purebred flock.

The White Rock, White Wyandotte, and Rhode Island White would appear to be good choices for this pursuit. The White Rock has long been known as the workhorse of small farm chicken breeds. We had a hatchery flock of them that laid measurably better than the breed norm, but they were of a smaller more "eggy" type.

By accepting that all pure breeds were developed for rather specific niches you accept what can and cannot be done with them. Within the parameters for which they were developed you should find much room for further development along several courses. Most do have a practical top dictated by how and to what ends they were initially developed. Producers can make Brahmas better layers or more meaty, but you can't make them Minorcas or Cornish.

Accept and actually revel in this, for then you are a student and a respecter of your breed or breeds. A good White Crested Blue Polish can continue as both a thing of beauty and a bird capable of earning its keep. It can produce a good sized white egg, appeal to 4-H'ers shopping for fancy showstock, be fairly efficient to own, appeal to your sense of aesthetics, and give you the challenge many breeders seek to keep them focused and interested.

I have moved some birds in and out of the keeper pens two and three times. My iron rule is the third time I question them they have to go, unless . . .

There is no ideal setup, no perfect form to follow in working with and propagating rare and heirloom chickens.

The key may be to find at what level of production you are most comfortable and at what level you can most profitably operate with those birds. Get those two reconciled and you've got her on a downhill slide. From fewer than ten hens you could produce enough eggs for the weekly needs of a family of four. From them could also be hatched enough broilers and roasters for the family freezer and replacements for next year's flock.

Buyer demand for various breeds can be somewhat difficult to determine and may sometimes be influenced by some rapidly changing fads and fancies. Some breeds experience fairly consistent levels of demand and others may be truly hard sells.

The heirloom breeds will be what we as individual producers make of them and no more. They can be lost entirely with but a very few wrong turns. Breed preservation work, if it is successful, will eliminate itself over time. They will be restored to wide acceptance and use. They will be all the way back!

Some were never kept in truly great numbers. And why you find a breed appealing may be for reasons that are all your own. If you are to be truly successful with them you must invest something of yourself in them.

You must own them with a plan of action that will build upon their and your strengths. You have to work at them every day and in all ways.

KEEPING A FEATHERED HERITAGE ALIVE

Dawn Turbyfill of Heirloom Heritage Farms in Spanaway, Washington is a seventh-generation farmer who's doing her part to save the face of endangered poultry. After spending much of her life in her native Wisconsin, four years ago Dawn and her family decided to buy a farm and move to Washington. Little did she know that what began as a simple desire for some eggs for the family would end up creating what is now one of the largest collections of rare poultry in the Pacific Northwest.

In the beginning, Dawn stumbled upon the American Livestock Breeds Conservancy (ALBC) while researching different chicken breeds. Through the ALBC she learned that there were breeds of domestic livestock that were disappearing at startlingly rapid rates, often as high as four per month. So then and there Dawn decided that if she was going to be raising chickens anyway, they might as well be worthwhile. Soon after her discovery, Dawn joined the ALBC and began purchasing poultry that fell under the "Critical" and "Threatened" lists. When Dawn first started Heirloom Heritage Farms she housed around ten chickens from only a few breeds, today that number has multiplied to 400 chickens from 40 different breeds varying from Barred Holland, White Sussex and White Orpington to Buckeye, Sicilian Buttercup, and many more.

When asked why she decided to focus on the preservation of poultry Dawn said that contrary to what many may believe, chickens are very intelligent creatures with complex and interesting personalities. Dawn said that the individual mannerisms and physical differences between each of her birds is what keeps her entertained and continually fascinated by the feathered livestock on her farm.

The strategy for breeding at Heirloom Heritage Farms is summed up by Dawn as a "numbers" game. Where other breeders may focus on strict guidelines for genetic aptitude, Dawn says that the goal on her farm is to maintain the breeds and hold onto the birds that live long lives and produce enough eggs to keep a breed thriving. According to Dawn, the poultry on her farm is meant to serve a purpose: to help the farmers. And if a breed isn't doing that, well then the Maintaining a rare poultry farm is no easy feat and Dawn will be the first to tell you that. With the sky-high prices of feed constantly on the rise and governmental interference in over-seas livestock trade and regulation, keeping up a farm full of endangered poultry is becoming increasingly difficult. So what keeps Dawn going? She says it's the look on people's faces when they see her birds and say, "Wow, I didn't know that this many different breeds of chickens even existed!"

Dawn's advice to anyone interested in getting into the business of rare poultry preservation is simple: be realistic. The possibility of getting rich quick from this type of work is slim and the farming industry is a constant uphill battle that will frustrate you to no end. But Dawn says that at the end of the day, if you can look at your birds and smile, if you can sit back and have a chat with a chick about what an exciting day it had in the coop, well, then that makes it all worth it.

Dawn Turbyfill raises a variety of rare poultry as well as sheep, goats, and alpaca on her family farm in Spanaway, Washington. She has given speeches in the Washington State Senate regarding the preservation of endangered poultry and welcomes visitors to her farm upon appointment.

Heirloom Heritage Farms is in Spanaway, Washington and Dawn Turbyfill can be reached at (253) 375-7444 or visit www. heirloomheritagefarms.com.

• CHAPTER 7 •

HENS, HATCHING EGGS & CHICKS

BLACK LANGSHAN
• MALE •

BLACK LANGSHAN
• FEMALE •

SILVER SEBRIGHT
• BANTAM MALE •

SILVER SEBRIGHT
• BANTAM FEMALE •

THE CHICKEN EGG IS TRULY A WONDROUS THING. IT IS A GEMSTONE OF SORTS IN WHICH THERE REALLY DOES EXIST THE SPARK OF LIFE.

We produce and set hatching eggs for as many as eight months out of each year. For a time we even sold several breeds of hatching eggs to a Midwestern hatchery. It has been quite a learning experience, one that still has a new lesson to teach from time to time.

An egg may not last long in clumsy hands, but it is actually a quite durable and forgiving bit of natural creation. Whether the egg is the chicken's way of getting another egg or the chicken, the egg's way of getting another egg, I will leave for philosophers to debate. To get the chick from the egg is what will be discussed here.

Most of my generation of poultry farmers began with hatching eggs the really, really old-fashioned way — under broody hens. For us it was the least costly option and this method was pretty much foolproof if the hens had what it takes to hold the nest and the critters left them alone.

Even today the setting hen still has a role to play on some smallholdings. She is the favored option for those working with some of the real exotic forms of fowl. Some old hands will trust their most needed and valued eggs only to proven broody hens.

The two big problems with broody hens are their somewhat limited capacity to cover eggs and the simple fact that they don't always go broody when you need or want baby chicks. I know of no way to make hens go broody although it has been said that if the trait is in them it can be sometimes be triggered by switching them to an all grain ration in warmer weather.

A clutch of hatching eggs traditionally measures fifteen, an unusual number even for our non-metric system. It is the near ideal number for a good-sized hen to fully cover and then brood successfully.

Some standard bred hens can have this number bumped up to about eighteen eggs and some bantams will do best with just eight to ten large eggs. The tendency to have a broody nature has now been bred out of a great many breeds and strains within breeds. Although I have heard reports of even the odd Leghorn going broody over the years, the trick now will be finding dependable setting hens. Most who use broody hens now maintain a special small flock that has been selectively bred just for this trait. There are a few Standard breed varieties offered as retaining the broody instinct, but wade into those waters rather carefully. If you do find a line in which the hens will take to the nest and hold to it consistently by all means propagate them. However, good setters are generally not high output layers.

One of the few good reasons for a smallholder to do much in the way of crossbreeding is to develop a small flock of dependable, broody hens. An acquaintance has a set of hens based largely on some old line Buff Orpingtons that will even hatch peafowl eggs and go on to brood and rear the peachicks. The best foundation birds for a flock of broodies may be best drawn from the bantam sector.

Silkies have a reputation for trying to hatch everything including doorknobs. I have seen Silkie hens go broody after laying as few as a half dozen eggs and they are truly staunch on the nest. They can handle up to twelve eggs and with a small flock kept just for broodies you can selectively breed for larger sized hens. Some will cross Silkies with other bantams to produce a somewhat more vigorous flock of broodies.

Silkies are very gentle, do not fly, and are the classic "floor" birds. They fall into a category of their own somewhere between bantams and standard birds although they are most often seen and shown with the bantams. Due to their truly gentle nature they should not be penned with some of the more aggressive breeds.

Some have opted to keep a small flock of purebred Silkies with which to do their hatching and maintain the purebred theme of their other flocks. They always sell well, too. Some will say that the only true Silkie is the White variety with its very dark, mulberry colored skin. It is a highly valued table bird in some Asian cultures and is also esteemed for certain medicinal properties. It is seen in a number of other colors and patterns than just white, but as I have handled them I have noticed that nearly all had at least a few hard feathers.

Cochin bantams have been used often for natural incubation, but with their feathered legs there can be problems with cleanliness and even eggs being pulled from the nest. They may be more useful for this purpose if crossed with a similar, clean legged, larger sized bantam such as the Wyandotte. Among the standard varieties some lines of Rocks,

Wyandottes, and Orpingtons retain broody tendencies. It is a trait that seems to manifest itself stronger in the rarer and minor strains within some of these breed categories. Blue Wyandottes and Buff Rocks may have been among the broodiest large breed birds we've owned, but birds from a different strain may not be nearly so strong to the nest. Standard Games often retain quite strong broody character and can be real chick raisers, but they can be quite aggressive toward other birds.

Before trusting a hen with a clutch of rare and very valuable eggs make certain that she will hold fast to her nest. To do this, test her with a nest egg or two for a minimum of forty-eight to seventy-two hours.

The containment for brooding hens should be some distance from other birds to keep the broodies calm and to prevent other hens from laying to their nest. A quiet, darkened area is best and many once built small, broody hen coops. They were, essentially, a two-wide stack of nests made of solid construction to a height of six or seven feet. Often they have a house-type door with some screening for natural ventilation.

The hens would be let out for an hour or so each day. The birds would be supervised and the nests then be checked for broken eggs, parasites, and soiled nesting material. Most now just partition off a darkened corner of a larger building or use a small outbuilding to contain broody hens shown to be solid to the nest. Such setting hens can be penned together if their nest sites are clearly established and they can get off the nests to feed and water themselves. Their nests will still need to be regularly inspected.

For best results broody hens should be healthy and free of parasites. Nest boxes should be size appropriate for the birds being employed. Nest size can range from twelve inches by twelve inches by eighteen inches to a full twenty-four inches by twenty-four inches by twenty-four inches.

Such nests will need some preparation before introducing the hen and eggs. It should have been cleaned thoroughly — including a scraping of all surfaces. Treat the surfaces with a miticide if there have been a previous problem with them. Old timers will apply a lime-based whitewash to all surfaces before each new use. Organic producers might want to look into using MiteGard for mite control.

Apply a layer of cedar chips or shavings to the bottom of the nest box. These contain a natural insecticide. Over them apply a layer of clean straw or other nesting material to a depth of two to four inches. As time passes this material may have to be supplemented and badly soiled spots removed carefully.

If given the needed space and protection, a broody hen can be entrusted to raise her chicks, also. This will require added time and facilities and will have a hen out of lay for many weeks. A great many prefer to pull the clutch of chicks once hatched and brood them in regular chick brooding units. This does give an added element of control over how the birds are fed and tended.

By removing the chicks it is possible to have a broody hen bring off two or even three clutches in a single season. Some will even have Silkie and Silkie-crosses incubate two clutches of eggs in succession. I once had a little Silkie hen take up residence on a bare nest on Labor Day and hold to it until Christmas Eve.

For those who want true and natural simplicity, the broody hen may still be the way to go. They lack a certain predictability as well as time and volume controls, but they do have longstanding history and old Mother Nature on their side.

INCUBATORS

The purchase of a cabinet incubator with a capacity of several hundred eggs was what marked the real launch of our work with rare and heirloom chicken breeds. Into that machine goes eggs of all sizes and colors and, if all goes well, out of it comes an equally wide array of baby chicks of nearly every hue and stripe. At the end of each hatching season and again the following spring, I give the unit a most thorough cleaning. I scrape down all of the surfaces and wash it out with a solution of bleach and water. It is a real roll-up your sleeves task as you stretch to reach into every corner. It is always amazing to me where little bits of eggshell find their way in its interior.

The clear door is washed, the door gasket checked carefully, egg trays and racks washed, and they and the humidifier tray set out to catch some sun. I try to expose the trays to the naturally disinfecting rays of the sun often during the course of the hatching season.

Along with a second spring cleaning, everything mechanical is given the once over. Thermostat wafers are examined and replaced (our unit has two), every surface is wiped down again, and the unit is turned on for a full trial run several days before it will be actually needed. The test run is to see that the turners work, temperatures are held, and that all racks are free.

Signs of thermostat wafer wear or failure will include a squashed or swollen appearance or surface discoloration. We use a digital thermometer to gauge operating temperature and like it much better than the dial-type with stem that came with the unit. It does take a bit of time to get operating temperatures right after a restart from a prolonged rest or when wafers have been replaced.

Around here that task sometimes takes a long afternoon, one in which my wife and the yard dogs just seem to know it is best to leave me completely alone. Ideally, an electronic thermostat would be best as they can be fine tuned in their settings, but it is not a stock part on some of the older units. It can add roughly one hundred dollars or so to the purchase price of a new unit.

In early spring especially, swings in barometric pressure can skew wafer setting and operation. Also, an incubator full of eggs will heat and hold temperature differently from an empty one. I generally try to have our unit up and running correctly for seventy-two to ninety-six hours before the first eggs go into it. The thermostat setting may then need tweaking once in use, but these occasions are rather few.

I know it doesn't sound very professional to say, but most years in Spring I just finally say that's enough and leave the thermostat alone after that first day. The other tool used in monitoring an incubator is a hygrometer, which is used for measuring humidity inside the incubator cabinet.

We don't own a hygrometer, but we are in the Midwest where summer humidity has never been a problem — unless you try to breathe in late July. Actually, cabinet humidity may be the most discussed and debated topic among incubator owners. Our unit has a small plastic, open-topped tank positioned above the eggs that we top up with warmed water every few days when it is in use. Wicking blocks can be placed in it that will pull added moisture into the path of the air-circulating fan behind it.

I am especially conscious of keeping the tray topped up when eggs go into the hatcher tray at the bottom of the cabinet on their eighteenth day of incubation. Too low of a humidity level and the chicks will have trouble hatching and dry too quickly. Too high and they may emerge, but die soon after, looking wet and swollen.

I have friends that will mist their eggs with warm water once each day. They will leave the mister bottle in the incubator on the shelf next to the humidifier tank. Conversely, I know others who will use no humidifying water at all. They depend solely on the Show Me state's fabled natural humidity — it has worked just fine for hens for a long, long time.

To better maintain the healthfulness of the cabinet environment, some will add a bit of chlorine bleach to the water in the humidifier tank. They add one teaspoon of bleach per pint of water and may do so once or twice a week. Too much chlorine may kill incubating embryos. This can be an especially valuable practice with waterfowl eggs that are often laid willy-nilly around the poultry yard in the damp and muddiness of early spring. A misting product called Oxine is also good for this process. It is OMRI listed.

If you do have eggs go bad inside the incubator and especially if one goes off "hand grenade" fashion, a full take down and clean up of the cabinet interior may be in order. The bacteria released can really take a toll on embryos.

Now for one of the things that worked for us, but might not for everyone else — we once unplugged a filled incubator, wrapped it in old blankets, moved it forty miles down the road, set it up in a new site, and still got more than a seventy percent hatch. Keeping the unit level and insulated can do much to bolster incubator performance. During a power failure, an incubator can be wrapped with wool blankets (hit up the surplus stores) or other insulating material such as a water heater cover and if the cabinet is kept closed, the eggs will be at least partially protected for several hours. The hatch rate will be reduced, but it will buy some time and help to save at least a portion of the hatch.

While on the subject of emergency management, always have on hand a backup kit. In this should be extra thermostat wafers, an extra thermometer, operating instructions, and other spare parts. With the cabinet model with which I am most familiar, the electrical component most apt to fail is the toggle switch that activates the turning mechanism. These are simple, three-position toggle switches.

With the incubator up and running, the next component of the successful hatching equation is the egg — "ovum" in Latin.

My great-nephew Kendal got a real thrill early in life when I put a near to hatching egg in his little hand. It actually turned in his hand from the inside chick's efforts to be born. He held it to his ear and heard soft peeping coming from within. With youthful resolve he set about learning all that he could about "hatch eggs." His great-aunt was sternly warned the first time he saw her putting surplus eggs in the refrigerator.

There is both a bit of science and a bit of art to the successful handling of "hatch eggs." Not every egg will hatch, not even every fertile one. The task is to keep them fertile, viable, and moving into the incubator in a timely fashion.

HANDLING EGGS FOR HATCHING

When gathering and handling eggs meant for hatching, there are several things to consider.

• Gather eggs to be hatched a minimum of two to three times a day. In cold weather gather the eggs a minimum of three times each day.

• Keep nests well bedded with straw or other soft, dry, and absorbent material. Giving the birds as clean an environment as possible into which to lay their eggs will both save labor and improve hatchability.

• In wet, muddy weather hold the birds indoors until at least 10:00 a.m. to get the eggs laid into dry nests. Provide added nest space if doing this.

• If eggs are lightly stained try to clean them with a light scraping using a dull blade or gentle buffing with fine grit sandpaper. Discard any badly stained eggs and those stained with the liquids from broken eggs.

• For best results store hatching eggs for no more than seven days. This figure can be pushed to ten days, with fourteen days about the maximum to push your luck. I've seen texts that report fair results with eggs held for up to twenty-one days, but then I've also heard stories of eggs pulled out of a refrigerator after many days and still hatching. I don't want to take such a chance nor own such a refrigerator.

• The best temperature level at which to store hatching eggs is in a range between fifty-five and sixty-five degrees F. In very warm weather I will hold hatching eggs in an all-wire rabbit cage, in an open incubator tray, and have it suspended in a shaded area. It generally takes us two to three days to fill such a forty-eight-egg tray.

• Store the eggs small end down.

• Store the eggs in cartons or flats of extra incubator trays. Elevate one side by placing it on a brick or empty egg carton. Then each time you pass them by, turn and elevate the opposite end. This will prevent the air cell from sticking or becoming mispositioned.

Fresh and clean are the two keys to successful hatching egg management. When hatching for yourself, eggs can go into the incubator at just about any time of the week as long as they won't later overtax

TAKEN OVER BY BABY POULTRY!

I have just a few minutes before having to go check on baby chicks and turkeys, refill water jugs, replenish feed trays, and check the heat lamps. You see, right about now, that's how we spend our days. We have so many babies on our farm and more on the way! Here's an inventory of what we are currently caring for (and this is just what is being brooded, never mind what's on pasture):

50 Narragansett turkeys
250 Poulet Rouge chicks
100 Bourbon Red turkeys

Coming next week:
50 Bourbon Red turkeys
250 Grey chicks

Now, about the chickens . . . we did not want to raise the traditional Cornish/Rock as our meat chickens and we are excited to have found the Golden Rangers. We have been pleased with these chickens and have enjoyed raising them, but unfortunately, the company we purchased the day-old-chicks from has gone out of business. We have already ordered and received 250 new broiler chicks though from another hatchery, and we feel that it was a blessing in disguise. We have been interested in the Label Rouge movement in France and wanted to raise breeds similar to this, hence the Golden Rangers. The naked-neck chickens were our real aim, but we just couldn't find them anywhere in America, until now! We found a supplier in Alabama who has taken the original strain of naked-neck chickens from France and simply bred them to put the feathers back on their necks because he found people liked the look of them better this way.

I say "simply" but actually he's a geneticist and I'm sure there was a lot of time put into this "simple" procedure. Anyway, after searching him out, we decided to try some of his "Poulet Rouge" or "Naked-Neck Red" chicks for our next batch of broilers. They are slower growing than the Cornish/Rocks, but still a relatively fast growing meat bird. Also, they range much more than the Cornish which means they will forage well and should grow steadily but not too fast for their own good. They are almost 3 weeks old and doing fantastic! I was so upset on our first batch of Golden Ranger chicks that we lost the traditional 5%, but so far with the Poulet Rouge we have only lost 2 — not 2%, just 2 chicks out of 250!! I can't wait to see them grow more and venture out to the pasture.

After many conversations with the geneticist in Alabama, we became intrigued with what he is doing. His interest seems to lie in taking traditional breeds of chickens and breeding them for characteristics that make them strong on pasture. In particular he began this venture because he saw a lack of supply in the free-range market place. He is able to retain the great "chickeny" taste of a traditional bird and the instincts of heritage breeds, but selectively breed them for faster growth on pasture. Before you begin to think that this is how the Cornish/Rock disaster started, let me say that throughout his breeding, one of the characteristics he values is that the chickens continue to act like chickens — they will reproduce on their own, they will forage for grass and insects, and they will know to get up away from the feeder and get a drink of water when it is hot outside!

We have been pleased with the Poulet Rouge chicks, but we decided to try yet another breed of chicks for our next batch because he has told us about some chickens that he has derived from Barred Rocks. He calls them Greys and says that they are of excellent quality. We figured that it would be a good idea to compare the breeds and get feedback on taste before settling on our favorite. We are very excited to receive them next week and are grateful for the relationship we have found with this hatchery. Their service has been great and their knowledge of pasture-raised poultry and free-ranging chicks has been very helpful to us.

Liz Young, posted on the Nature's Harmony Farm website www.naturesharmonyfarm.com.

brooding and growing facilities. Hatcheries generally set eggs to hatch on a Sunday or Monday to get the chicks shipped and arriving at their new owners by early to mid-week. Producers can likewise schedule hatches to arrive in a timely manner for growing out birds for show or to arrive for specific marketing events such as farmers' markets.

For all of the best efforts there also seems to be that element of luck needed to bring off a good hatch from time to time. Even the definition of "good hatch" can be something a bit different with the rare and exotic breeds. With White Rocks in big cabinet incubators, we often approached one hundred percent hatches. With some of the rarer breeds and those from small, closely related populations, a fifty-five percent hatch may be something to crow about. Sixty-five to seventy-five percent hatches with a smaller incubation unit are always quite respectable.

Hatchability is the first proof of a mating. Clear eggs, those that were not fertilized, will result in a need to reevaluate your breeding birds, the weather, ration choices, and a number of other things. Older roosters will generally perform later in the breeding season. A rooster much over four years old must always be considered a bit suspect and especially so with the larger, heavier breeds.

Keep spare males on hand, should performance from a breeding pen falter or a male be lost. A younger male may be all that is needed to get a set of hens back to producing fertile eggs.

Another one of those much-discussed subjects among poultry raisers is how best to ready eggs to go into the incubator. Everyone seems to have their own list of do's and don'ts and what works for one person may not work every time for another. One of the most controversial practices was actually taught to me by a commercial hatchery operator. He wanted all eggs delivered to him to be dipped in warm water to which a bit of bleach had been added and then they were to be wiped clean with a soft cloth. This water was around 101 degrees F. and we would add about a tablespoon of bleach to two quarts of the warmed water. This practice is a real anathema to many, but this man regularly reported ninety percent plus hatches from our washed eggs.

I don't wash eggs often, but will if it keeps me from having to discard some very rare or valuable eggs.

INCREASING HATCHABILITY

Setting day is also an important one when hatching eggs and includes several point to increase hatchability.

• Be mindful of the day the eggs will hatch. Don't set them to hatch on holidays, times when you will be away or to arrive in busy seasons on the farm.

• Many will candle the eggs before placing them in the incubator. They are looking for fine cracks and other structural anomalies. Do not incubate overly large, double yolked, misshapen, rough shelled or small and pullet eggs. As noted earlier I have violated this pullet egg caution and especially when trying to build numbers of a really rare variety.

• Maintain a careful log as to the incubator's loading and operation. Document when eggs go in, where they are positioned in the unit, the breeding behind them, and when they are due to hatch.

The incubator is viewed by some as a strange and even mysterious thing and to others it is just one more tool to be used daily. On more than one occasion our incubator has confounded me, but what becomes clear over time is that it is rather like owning an old pickup. Things just seemed to run better once it and I got attuned to each other's quirks.

On the eighteenth day of incubation, eggs should be removed to the hatching tray of the incubator or to a separate hatcher unit. Many do not like to hatch within their primary incubator believing that chick fluff and hatching wastes can build up and affect components or contaminate the cabinet environment.

We have used the hatching tray in the bottom of our incubator for several years now and continue to have good results. We remove it several times each season, scrape and disinfect it, and expose it to the rays of the sun. Still, a separate hatcher is one of those things on our someday wish list. Before placing the eggs in the hatcher, we candle each one and remove those that are not bearing a viable embryo. Handle these blanks carefully as they can go boom on you. I take these eggs some distance away from the house and break them just to see how they did or didn't develop.

Some will candle eggs multiple times during incubation, including around the seventh and fourteenth days. I don't always trust my old eyes for an early candling and prefer to keep my entering the incubator and handling of the eggs to a minimum.

In the hatcher, when the eggshells break away, the background identity of the chicks written on the shell can be lost unless some specific measures to safeguard them are taken. As the eggs are candled on the eighteenth day they can be positioned in the hatcher in certain ways that will then enable you to pinpoint the origins of newly hatched chicks.

IDENTIFY HATCHING CHICKS

There are several ways to accurately identify chicks as they hatch.

At one time, eggs close to hatching were actually placed in rather elaborate drawstring pouches. A modern version of this is to use simple bags made from the legs of old pantyhose. One end can be knotted. The egg is placed in it in a manner that leaves the emerging chicks plenty of flex and the other end knotted shut. These securely separated eggs can then be placed in the hatcher. The hatched chicks are then removed from these porous containers that are then discarded.

I have also seen hatcher trays that were segmented with thin wooden or wire mesh dividers. Eggs from different matings are placed in the different segments and their location in the tray written into the incubator log. The hatcher tray top then holds the newly hatched chicks in position until they can be toe punched or otherwise identified. A dot of nail polish has been used as a very short-term method to denote different chicks.

Small, pedigree-mating baskets can be made from fine wire or plastic mesh. In them the eggs are completely contained, as are the chicks that hatch from them. Some have even cut down plastic berry boxes for this purpose.

I have seen bantam chicks hatch as early as the nineteenth day of incubation. Some of the very large breeds are tightly bred birds and may continue to hatch into the twenty-third day. Ideally the chicks should remain in the incubator for several hours after the last chick hatches. I may pull chicks twice from the hatcher and when lots begin to back up have even placed a few still damp chicks into a brooder. I try to keep my racks of incubating eggs at least three days apart and prefer four or five. When egg production is at a peak and I am trying to get maximum production from some of the rarer birds, this rule sometimes gets bent.

A tough call to make is whether or not to help a chick that is having a hard time emerging from the shell. A first step might be to mist them a bit with warm water or even dip them quickly into the humidifier pan if they appear dry. If they appear wet and bulbous, the problem is too much humidity.

Just how much to help is the real question. I have helped to make an initial pipped opening larger or even pick away some shell pieces and have had a bit under a forty percent success rate with helping late hatchers. These measures should be done very carefully and only by removing very small areas of shell. If blood should appear stop immediately and put the egg back in the hatcher, but sadly, if that happens, too much damage has probably already been done. Where absolutely every chick counts you will have to rush the envelope, but those slow hatchers could still come back to haunt you by producing more of the same kind.

As noted earlier, the traditional hatching season in the United States runs roughly from February through May. Chicks hatched then will reach laying age sometime in the fall. Meat birds should attain a good harvest weight in the traditional, barbecue/frying chicken summer season.

Our hatching season tends to run longer, into early fall, as we try to maximize production from some of our rarer breeds and breeds that didn't perform well earlier in the year. Some breeds such as the Sumatra just do not reach peak fertility until the warmer months of the year.

We too often seem to overmarket our early-hatched birds and have had to scramble late in the season to get the needed numbers for the next breeding season. These late-hatched chicks do seem to find a ready market in the first warm days of the following year as people start getting "chicken fever" again. There may be no more valuable product than ready-to-lay pullets and young breeding birds in February, March and April.

Our incubator holds six, forty-eight egg trays on its three racks and I believe it performs best when full or nearly so. We have also learned to expect lower hatching percentages late and early in the year and following prolonged hot spells. If the heat winds down here in late August, our hatches in the second half of September may really rebound.

Fertility and hatchability fit together like hand and glove. Producers may need to take special pains in the care of the breeding flock to achieve good hatchability. We recently went through about a six-week spell of absolutely no production of chicks from a rather large group of Welsummers. It was a first of sorts here and a good deal my fault as I was overly enamored of the Welsummer male that headed up that pen. I forgot my own admonition against becoming "bird blind." The bird demonstrated all sorts of vigor and libido, and sterile males in that breed had been almost nonexistent in my experience.

That rooster certainly broadened my experience, but he wasn't totally at fault. After he was replaced with another, proven male roughly a full third of the eggs from that pen still failed to fill. That pen was soon broken up and nothing retained from it and we chalked them up to just one more signpost on the old learning curve.

When hatchability isn't what it is hoped for, the first place to look is the breeding pen. Has there been heat stress? Has a parasite load gone undetected? Are they simply not up to the task due to age or nutrition? I will accept a lower level of production from older birds if the birds I am getting are good ones.

The place we normally begin is to boost protein levels in the breeder rations. We'll generally use fish-based cat food or game bird layer to accomplish this simply and quickly. It is also the time to take each bird in hand for a thorough examination of health and overall condition.

Also, strive to be ever more savvy about the poultry scene and what goes on there. Last year I saw offered for sale a set of Black Cochins — really big and showy. Fortunately, they belonged to an honest man. Those huge and docile males were from birds that had been produced for years via artificial insemination. They were just too large and too dead headed to any longer be successful at natural service.

With some of the larger and softer feathered birds there can also be breeding problems. The coupling is sometimes incomplete and blocked by the feathering. To counter this it may be necessary to pull away some of the feathers surrounding the vents of both the male and female birds. Pull, but do not cut or clip these away as this can leave a hard, sharp feather stubble that can also deter mating behavior. With rare birds and breeds from small populations producers must make size and vigor absolute musts in the selection process. No matter how pretty they are, no one can afford a rooster lacking in rattle and snap. And those little and cute pullet chicks generally make little and poor performing hens.

Many years ago there was a very popular book entitled *The Egg and I*. What I have long wished for is a handy little primer entitled "The Chick and You." When I pull that tray full of newly hatched peeps from the incubator I am as giddy as a kid at Christmas and as optimistic as a teenaged boy. They're all mine and they're all good. Well, not really.

There are some obvious problem chicks to be eliminated right away and these include chicks with crossed beaks, spraddled legs or umbilical ills. Other problem birds will emerge as time passes, but begin by taking the time to closely examine each chick before it goes into the brooder.

The small-scale producer with heritage breeds will be producing for two distinct cycles of production. The first is generally termed the "numbers building" phase. Here just about every egg laid goes into the incubator and virtually every chick hatched gets the test of time to prove itself. The one big question in evaluating them is, "Is there anything here at all with which to work?"

The Exchequer Leghorns have been one of the biggest challenges I've faced recently. They're fair layers for their category and they're fertile, but the chicks could have more vigor. Other challenges with them have been size, leg color, and feather color pattern. We still get too many smallish birds, pale legs, and birds with just too much black coloring.

We are also flying a bit blind here as this is a breed currently outside of the United States Standard and pictures of good specimens are few and far between. What we have now aren't exactly what I remember being depicted in the catalogs of my youth. A few years back some chicks were held back that were largely white, but each feather development stage saw them grow progressively blacker. Bear in mind that many breeds, especially the larger ones, are not fully developed until one year of age and some traits take longer to develop than others.

Our Rosecomb Rhode Island Reds, for example, are pure in their breeding and quite dark red in their color. At a distance they can appear black. Many dark feathered birds are quite slow to feather out and can look quite ragged at even eight to nine weeks of age. Some of the better young cockerels have thin body feathering for many weeks after their neck and wings are feathered.

Certainly, you want to cull undesirable birds when they are as young as possible to save on both space and feed needs. During the numbers building stage it can certainly help to pare costs by having a breeding partner or two. Producers can "farm out" some of the chicks to a breeding partner and the different flocks might even then key on developing and improving different traits.

With viable numbers established, the second level of production is meant to maintain the home flock and to provide the numbers needed to serve any and all markets that have developed. This corner is a firm believer in growing numbers to meet a market's demand rather than producing heavily and then venturing out into the world and asking, "What will you give me?"

Most will need to replace five to twenty percent of the birds in a rare breed breeding flock yearly. Some will opt to completely change a laying flock each year and to that end will need to start 20 to 30 percent more than the number of female birds ultimately desired. With a large flock this can mean hatching several hundred chicks just to maintain numbers.

We are comfortably operating with a hatch of twenty-five to fifty chicks a week in total from the breeds with which we are working now. It is our hope to gradually increase these numbers, but we will take our cues on this from those that are buying from us. Their input has already altered our mix of breeds and has us fixing on the more practical rather than the more ornamental varieties.

Our Reds and Orpingtons will, no doubt, always outsell the majority of our other breeds. We simply cannot dictate others' wants and just work to make their and our wishes merge as best we can. Should you be fortunate enough to be working with a breed that catches fire, your sales will certainly spike for at least a short time. However, don't base your long term planning on those prices lasting nor use their popularity to dump poorer quality stock on the unsuspecting and the inexperienced.

Only time and experience will tell producers exactly to what levels of production they should strive to reach. Experience over time will also be the guide as to how to plan a hatching season. As producers reach beyond local sales they may have to hatch earlier or later in the year or both.

My friend with the Braggs Mountain Buff breeding flock has had a real learning experience in supplying chicks for a national market. A few want chicks as early in the year as possible and this will be especially true for those wanting birds to exhibit. Some in areas of climate extremes factor this into their order dates. An early order that pans out well may result in one or two more later in the year.

After the how-to's of laying flock culling, the most common question I am asked is how to sex young birds. Sexing day-old chicks is a skill that has been mastered by but a handful of people in this country. Several of them travel from hatchery to hatchery in the spring and early summer doing such work for up to twelve or fourteen hours each day.

There are texts that illustrate what to look for when vent sexing baby chicks, but this is no simple task and the visual indications are often quite slight. Many will find it difficult to bring themselves to handle baby chicks in the way required to successfully carry out this task.

There are a number of visual clues that many use to sex their own birds. These are oftentimes little more than folklore and you certainly won't use them to guarantee chick sexes to buyers. Still they may have some merit and are worth considering — just don't use them to cull through rare and valuable heritage birds too quickly. Prove them to your satisfaction before relying upon them.

VISUAL CLUES FOR SEXING BABY CHICKS?

There are many theories about the possible visual clues for sexing baby chicks. Some are more reliable than others. Good luck!

• The spot atop the head of barred birds is generally more round on females and more elongated on males. Females are also generally darker than males in their barring.

• An old adage holds that if a young chick is picked up just behind the head with thumb and forefinger and raised, a young male will struggle and a female hang limply. The same has been said if the chicks are suspended upside down by their feet.

• The legs and toes of young cockerels are said to be longer and larger than on newly hatched pullets.

• Pullets generally feather faster than males and especially along the forward edges of the wings.

• On Rosecomb birds, the combs of the males will appear wider and brighter than on the females

• Silkies are an especially difficult breed to sex and some will even wait until the males start crowing to be absolutely sure. A rule of thumb holds that if the crest feathers are spread apart, the young pullet's comb will appear more triangular and indented.

• Pointed eggs are said to produce cockerels and more rounded ones, pullets.

The most venerable of sexing devices, once even used over the abdomens of expectant women, was a piece of metal suspended at the end of a bit of string. If it swings back and forth it means one gender and for the other it turns in a circle.

To follow the letter of ancient lore it should be a silver needle suspended by a bit of red silk thread. Some of the above are a bit more proven than others and some are probably best as meat for coffee shop debates.

I did hear of a new use for the remote control from the TV as a chick-sexing device. Begin by inserting fresh batteries and place the remote in the brooder with the little guys. The chicks that gather around it and peck at the keys are little males. Those pushed back into the corners are the little females.

The young bird of the year is the ultimate product of the heirloom flock. They are the yardstick by which the year's work and the flock's progress are to be measured. In the early going with a very rare breed, success may mean you now have four females where you had just two the year before. Just adding a second breeding trio can sometimes be a real victory — a very big one.

Building on quality is one of those forever sorts of things. I would like to think that someday we will profitably produce and market five hundred to one thousand chicks and young birds yearly from each of our breeds. Will we?

Long-term numbers look good, but the actual production and breeding in the quality needed to create such demand will be entirely the work of our hands. What we have now isn't the breed mix we had even five years ago. A breed or two may be different in the mix five years from now.

Yes, I think narrowing our focus onto fewer breeds is the right idea, but it must never blind us to what is going on in poultry circles. No breed should be left behind. They all have history and tradition solidly on their side.

Five or ten years from now a White Wyandotte line could emerge with egg performance on a par with today's better White Rocks. Will there emerge in the future a super, meat-type Delaware or New Hampshire Red? What eventually will be done with the Light Sussex or the Red Dorking? Where is the modern niche for all of these traditional farmers' fowl?

Would the development of a good laying strain of White Crested Blue Polish lead to a single flock selling five hundred, one thousand or even two thousand as-hatched chicks per year for five dollars per chick? Answer questions like these and you will have a good handle on the heritage bird scene today, tomorrow, and for many years to come.

When breeding from small numbers and multiple matings you may encounter problems with hatching out and having to brood small and odd lots of chicks. Those quints and quads can be bad enough, but a time or two we've had to deal with an only child.

Chicks within just a few days of each other in age can be brooded together. Still, it can be a tricky call to make. Seven to ten days apart isn't that big of a gap if you're talking about three and four week old chicks. It is a chasm if one group is ten days old and the other newly hatched.

Some of the more docile and/or heavily feathered breeds should not be brooded together or grown out with other more vigorous or aggressive varieties. For example, a bird with a real good crest will have a most limited field of vision and it will be almost entirely forward. They are helpless from the back and some of the more elaborate feathering can induce feather picking.

Feather picking can lead to near-cannibalism, as the birds will bore right down into the flesh as soon as a drop or two of blood emerges. Remove both the victims and the perpetrators of feather picking if you can detect them. To detect the latter, look for traces of blood on the beak and head and the feathers of the breast. There are a number of anti-pick solutions that can be applied to birds that are being pecked and a salve-like product like petroleum jelly or Vicks will work in a pinch. Apply it heavily, as it is getting a beak full or two that is the deterrent.

Still it is always best to remove any birds that are being picked. A red tinted brooder lamp is also said to reduce picking behavior. Altering the beaks of aggressive chicks with a slight trimming of the upper beak with a pair of nail clippers may help some too.

Once past the downy stage things get simpler, but dampness and chilling can still be a problem, along with predators. From two weeks onward we work the birds at two to four week intervals and begin working on everything from color to crooked keels. Producers can't sell chicks with obvious defects, but those with color and simple type defects might suit those wanting just a few birds for the table or for backyard layers.

Birds being grown out for seedstock need to be culled quite closely and as the years go by with a heritage flock, producers need to be ever more exacting in the culling process. After four weeks of age or so some people will even separate the pullets from the cockerels.

We have had problems with dominant cockerels repressing the development of the other young males in the group. These are the males to select your replacements from, but remove them at an early age to allow the other younger males to develop apace.

The breeds we have now perform well when penned together and we may even winter groups of twenty or more replacements together in one group if they are close in age. We keep them on a good starter/grower feed until the first pullet eggs begin arriving.

Hatching, brooding, and grow out are all yearly tasks in the course of poultry flock maintenance and development. The numbers needed from year to year may vary. Also with different breeds on premises each breed-based flock may well be in a different state of development.

A HATCHERY THREE GENERATIONS STRONG

Cackle Hatchery is a third generation, family-run business located in Lebanon, Missouri. Clifford and Lena Smith started the business in 1936. With a strong family involvement, Nancy and Clifton Smith are the second generation and Jeff and Edie Smith, the third generation owners of the business that has grown steadily for over 70 years to now offer 181 poultry varieties to individuals, farmers and feed stores across the country. Jeff Smith calls the family business a hobby hatchery because they have a large selection of chicken breeds and serve both the wholesale feed stores and individual poultry keepers. Cackle Hatchery may ship an order of 1,000 birds for a feed store or as few as 15 for a backyard hobbyist.

The hatching season runs from February through September. Most of their customers want chicks in the spring to raise into young birds through the summer. Many want birds for 4-H projects that must be a specific age to enter competitions. With the demand for poultry highest in the spring, Cackle Hatchery does most of its business during the months of April and May. Once the calendar progresses into June, owner Jeff Smith says that the pace slows a bit and there's a quicker turnaround for processing orders. Cackle Hatchery is a seasonal business that does not hatch year round.

Jeff Smith offers some tips to help customers get the birds they want, when they want them. He encourages customers to order as early as possible. Orders are accepted after January 1 for shipment later that year. Orders received first are processed first.

Early in the year, an availability of breeds is created based on the previous year's sales. Anticipated increases or decreases in flock size and the average hatchability rates are also taken into consideration. Still, there is a lot of guesswork involved in predicting what varieties might be available and when.

Once orders are received, an estimated shipping date is assigned to it based on the anticipated availability. Even with all the possible glitches, Cackle Hatchery has an impressive 90-95% "deliver on the promised date" record — this, even for orders received 3 months in advance of the shipping date. Cackle Hatchery processes thousands of orders and strives to completely fill all orders as promised. Once in a while though, a few orders cannot be filled due to unforeseen circumstances.

Some types of breeds also sell out quickly and availability depends on the type and time of year. Some of the most popular breeds should be ordered three months in advance. Jeff recommends that for birds being delivered before June 1, orders be placed at least 6 weeks in advance. For birds delivered after June 1, he suggests ordering 2-6 weeks before the desired delivery date. Cackle Hatchery breeds their own stock with supply from their own qualified and inspected farms. Flock owners raise their own chicks and sell eggs back to the hatchery. At present about 63 flock owners in the Lebanon, Missouri area are under contract to supply a weekly allotment of eggs that Cackle Hatchery will use to produce the chicks they sell.

Cackle Hatchery specializes in supplying poultry for small farmers who would like some laying hens or chickens to butcher. Young people in 4-H and FFA are also frequent customers looking for birds to show or for pets. Wholesale orders for feed stores and similar rural outlets account for approximately 40% of the business. The other 60% of business is to individual retail customers. The vast majority is shipped through the postal system. Cackle Hatchery ships about 100,000 chicks each week during the peak of hatching season.

A typical hatching cycle at Cackle Hatchery starts with an egg placed in one of their 35 incubators for a 21-day incubation period. Starting at 3 a.m. on the day of hatching, the chicks are taken out of the hatchers for customer orders. Throughout the night and day they are counted, sorted by sex or breed and packed for shipping. By 6 p.m. that day, the last order is on a truck heading to the post office. Birds are shipped priority mail to their new homes throughout the United States.

The Hatchery has many customers that know that they want and have done advance research to select the breeds that fit what their farms best. The Cackle Hatchery website describes each breed and frequently has egg, chick, hen and rooster pictures too. An option is available to select substitutions at the time of ordering, but some do not go this route. Most prefer to stick with their first picks and will even wait longer to receive the breeds they want. If one type is not available in an assorted order, then usually the order is shipped only when the entire order can be filled completely. A good number of poultry are ordered online and shipped at the earliest date when all are available. Most customers are more than willing to wait to get all the birds they want.

A typical order might contain 25 Rhode Island Reds, 25 Barred Plymouth Rocks, 15 Silkie Bantams, 5 Rouen ducks and 7 turkeys. Still other customers will order 100 Cornish Cross in the spring and then arrange for some fancy breed pullets to arrive 6 weeks later once their pens are open again. It is common for multiple orders to be shipped to the same address throughout the year.

Many customers order by putting together their own assortment of 4-6 breeds, as opposed to picking a hatchery assortment. This customized ordering is a great opportunity for small flock owners to diversify their holdings. The "fancy birds" are what many hobby farmers want — after they get their egg layers and meat birds. Those that don't have a preference though, can always order the various hatchery-selected assortments that Cackle Hatchery puts together like the Frypan or Hungry Man Special.

The backyard poultry trend has brought many new customers to Cackle Hatchery. Jeff Smith shared an issue he has observed from some of these new customers. These new poultry raisers are highly fluent with online purchasing and come to Cackle Hatchery with an expectation that they can place their order electronically one day and have it shipped out the next without understanding about the poultry raising process and all that happens before chicks are shipped out the door. Cackle Hatchery has addressed this need by putting a "frequently asked questions" section on their website as well as including links to other websites that outline the poultry raising process and expectations. Cackle Hatchery also has extensive information online about chick care for new poultry owners to use before and after they receive their fluffy new additions.

Jeff Smith has also noticed more interest lately in heritage and rare bloodline breeds that farmers can reproduce themselves. More and more people are looking for good foragers, poultry that have long bloodlines and have been around a good while. Lately, many are getting away from the modern breeds that have been solely produced for excessive egg or meat production. Jeff Smith has observed that customers are becoming more educated now and are looking for a "back to the basics" breed of poultry and heritage breeds are what they are asking for.

When asked about lesser-known treasures, Jeff Smith sees great potential in the Marans, Welsummers and the dark brown egg producers among the heritage breeds. These he sees as being underutilized at present.

The debate about which is better — white or brown eggs shows no sign of diminishing. Jeff Smith offered his thoughts — he perceives brown eggs as having a little better shell quality when compared to white eggs. When asked about the public's views about which one is better, he commented that people are used to seeing white eggs in the grocery stores and think of brown eggs as being farm eggs. Consequently white-shelled eggs are looked upon as being mass-produced while brown eggs are perceived as being small farm raised with plentiful sunshine, free ranging, etc. Curiously, this perception is projected to the taste and quality of the egg also. So to many, the brown-shelled egg is perceived as being better tasting and of a higher quality. Jeff noted that taste and quality has much more to do with what the chicken is fed and not due to the color of the egg.

The White and Brown Leghorns are breeds that Jeff Smith recommends, that do produce white eggs. The egg color bias has to be overcome by some. Jeff remarks that Leghorns are good value for producers because they lay a good size egg, are efficient with their feed requirements and are an all round hard to beat bird. Their only downside (not counting their egg color) is their nervousness, as they are flightier than other breeds.

Alternatively, there are lots of good brown egg layers that are not as nervous as the Leghorns and that also have a good egg size. Jeff suggested the modern Cinnamon Queen and the heritage breeds; Rhode Island Reds, Barred Plymouth Rocks, White Plymouth Rocks, or Buff Orpingtons are also to be considered.

Cackle Hatchery used to offer hatching eggs but discontinued this because of shipping problems when sent through the post office. When there was breakage it was difficult to file a claim. Also during the peak demand time for hatching eggs, Cackle Hatchery has a big demand for chicks and limited supply of eggs to sell. Jeff said that they might try hatching eggs again but only for a few of the most plentiful breeds and only during May and June. This way Cackle Hatchery can use all the eggs during the earlier months for chick sales. They also don't want to sell hatching eggs in July because of the potential heat problems with shipping and the reduction of fertility during this time of year.

The heritage meat bird breeds can't compete with the excessive size and weights of the Jumbo Cornish

Cross, but Jeff Smith did suggest that the Rhode Island Red or Buff Orpington males did get to be a good size, they just take longer. Just because they are not quite as plump and juicy and some may be stringier than the Cornish Crosses, that doesn't mean that some heritage breeds are not an acceptable meat bird. Jeff considered the possibilities of a Red Cornish crossed with a Rhode Island Red (a good egg and meat producer) as a future meat bird offering at Cackle Hatchery.

The future looks strong for heritage birds. Jeff Smith remarked that there is a "big demand for heritage turkeys this year." He has noticed that consumers are more conscious of what they are eating. There is more interest in how food is being raised, with a desire to eat poultry that has been properly taken care of and includes the benefits of being raised in a pasture environment with sun and access to an abundance of minerals from the soil. These same informed consumers want to know what their birds were fed and to be assured that they were organically raised. Raising their own food is the logical answer for many people. Naturally, the heritage breeds fit this demand very well.

Cackle Hatchery is located in Lebanon, Missouri and can be reached by calling 417-532-4581, email: cacklehatchery@ cacklehatchery.com *or visiting their website at* www.cackle hatchery.com.

CHAPTER 8

MARKETING YOUR POULTRY

BLACK JAVA
• MALE •

BLACK JAVA
• FEMALE •

PARTRIDGE COCHIN
• MALE •

PARTRIDGE COCHIN
• FEMALE •

IN RECENT TIMES HEIRLOOM POULTRY PRODUCTION HAS BEEN SWEPT BY SOME QUITE MAJOR BREED FADS. The Dominique may have been the bird that launched it all and it continues among the most popular of all of the heritage breeds. Others that have moved up the popularity ladder include the Sumatra, Welsummer, and Maran. Currently on the upswing are the Buckeye, Laced Red Wyandotte, and Exchequer Leghorn breeds.

Why have these risen up in popularity? The Dominique had history, strong promoters, and a strong breeders' group on its side. The Welsummer and Maran built on the current interest in very dark brown eggs. That trend still continues now with the emergence of the Spanish Penedesenca. The Sumatra began drawing attention with its eye-catching black appearance and vigorous nature just a few years ago and now is also being bred in white and blue varieties.

With any livestock species there is always an element of great risk to fad chasing and the growing interest in rare chicken breeds is already being compared unfavorably with the exotic animal and bird fads of a few years ago. Yes, there are people who are in this business just to make money by jumping from one hot breed to another. I saw something similar to this several years ago with wild swings in popular swine type and the emergence of a new "hot boar" with nearly every show season.

Is the Blue Laced Red Wyandotte any better bird than the Blue, White or Partridge Wyandotte? What of the Buff Columbian and Blue Partridge Wyandottes that are found now only in Great Britain? They all have value and each will appeal to a good many different people. Their real value, however, will only emerge over time. When birds are produced with more than mere surface gloss and hyperbole, then real preservation work is being done.

Set up a crate of ready-to-lay red pullets at a Midwestern farmers' market on a bright April morning and they'll bring seven to ten dollars each. A trio of White Silkies of nearly any age is at least a sure twenty-dollar bill exchange at that same market.

That's at a local market. I'm not long back from a major farm show where I saw a couple of trios of Rosecomb Rhode Island Whites priced at $125.00 per trio sell to one buyer. A good friend has often paid five to ten dollars each for some of the hatching eggs she has used to begin rare breed flocks.

It is high prices like these that have fanned some of the burgeoning interest in heirloom chicken breeds. However, price is just one factor. Something similar is seen with organic, dark brown eggs selling for at least five times the going price for white eggs from caged layers. Likewise, the more than two-dollar per pound price for a range broiler has people thinking about chickens as never before in my lifetime. The range broiler position on family farms will be securely re-anchored only when consistent earning power from them replaces these rather spectacular, albeit short-term, gains that we are seeing now.

Some producers are seeing these spectacular prices plateau if not even falter a bit. Though Missouri law may allow it, no one now is planning to sell twenty thousand dressed range broilers any time soon, if ever. With an average dressed weight of 3.5 pounds and selling for two dollars a pound, they would produce yearly gross sales of one hundred forty thousand dollars. However, it would come at a most dramatic life-altering, farm altering cost.

A good marketing plan helps to cope with fluctuating markets, contains price-optimizing strategies, and gives the producer greater control over, rather than dependence upon, the selling price.

The traditional role of poultry production on the small farm has been to bring regular income onto the farm. This involves giving the farm family more to sell and a longer market across most weeks of the year. An investment in a poultry venture can literally begin generating some income within a matter of weeks.

A flock being set up to sell seedstock will take the longest to establish and then become fully productive. To be successful, all seedstock, meat birds or laying flock will require an investment in the marketing process as great if not greater than the investment in the actual production of the birds and/or eggs.

This corner has long felt that it is wrong on several levels for the small farmer to rely on all of his or her income from just one or two ventures. I dare say that now it would be all but impossible to create a single farming venture with the capacity to earn twenty thousand dollars a year in net income.

Ten small ventures each netting two thousand dollars per year would be far easier to accomplish. They should be varied, not directly compete for producer time and farm resources, give earning power throughout the year, and complement each other in overall tone. With such goals, a great number of such ventures can be stacked on even the smallest of acreages.

Should one or two falter or not fully suit the producer they can be folded without compromising the total farming operation. The ventures can be somewhat related or even grow out of each other, but no theme should ever become so oversized as to totally dominate the farming operation. We are now selling seedstock, table eggs, and a few specialty birds (pigeons and turkeys), but a group of five or more totally poultry themed ventures might possibly skew our farm mix too far in one direction. I realize that this flies in the face of the widespread belief that farmers now have to be specialists in one area or another. But really, where has that kind of thinking gotten us? If circumstances allow a certain venture to grow to considerable size, so be it. However, individual ventures should never be allowed to grow to the point where the producer is feeling pressured or overwhelmed. We frame this belief by thinking of our small seedstock venture as a micro-hatchery, borrowing from today's concepts of artisanal micro-vineyards and breweries.

The artisanal approach to production agriculture is one that, I believe, should be emulated more and more by independent farm producers involved in all types of production. They are about producing a distinctive product of superior quality and marketing it directly to those most likely to be appreciative of it and willing to pay a premium price for it.

In wine production you now see a lot of three digit bottle prices. Sometimes all three digits are to the left of the decimal point and sometimes two are to the right. What poultry producers' marketing plans should be all about is keeping as many of those digits as possible to the left of that decimal point.

Let us begin by discussing the marketing of the most basic of poultry products — the table egg. It is truly a staple of the American diet and may be the one single item most closely associated with the traditional American breakfast. Nutritionally, all eggs are the same and diet-wise, most people can safely enjoy one or two eggs each day. A great many Americans actually enjoy a "breakfast" meal at any time of the day. One of our favorite winter evening meals is scrambled eggs, sausage, and a biscuit. Our area produces all of the basics for this meal.

The English language is full of all sorts of poultry and egg-themed expressions. Some can help you, such as "The freshest eggs are in the country." Others can hurt you; "Eggs are cheapest in the country."

Getting eggs directly from the producer assures the consumer of the freshest table eggs possible. The typical store egg is now believed to be something like thirty days old at the time of sale. There have been food scandals based on the repackaging of eggs that had remained on store shelves past their "sell by" date.

The egg is a near perfect food, deficient in but a very few vitamins. Eggs lie in the hand smooth and rounded like a lump of jade. What else can be done to make this American food icon even more marketable?

The commercial sector pretty well boiled their product down to just medium and large white eggs a bit over a half century ago. Decades later their next big thing was to break them, dump them into one-gallon cartons and give the fast food trade "pourable" eggs. How technologically advanced is it really to break an egg?

Evidently breaking eggs is beyond the skill set in the fast food industry as a number of "egg-breaking plants" have been opened across the Midwest. This does create the ultimate, generic egg and that may indeed be its primary purpose. Pourable eggs may add to shipping and handling ease, but will any family farmer really be served by the marketing of egg slurry?

For most consumers the eggshell and its color are still very important to them. For some it may bespeak freshness, to others a more natural egg, and to some a more rare and desirable egg.

The concept of value-added agricultural production has been really encapsulated in the production of the table egg. We have come to a time when a large segment of the consuming public is acutely environmentally and socially conscious. This has led to many people with growing health concerns about what goes onto their tables now. These concerns are now being expressed in their buying patterns.

It cannot be helped but note with a degree of irony some of the promotions now being seen in certain egg marketing practices. Laying hens were among the very first livestock species taken into confinement and now a premium is being paid for humanely produced, loose housed, and free-range eggs. Further producer add-ons to attract consumer appeal include organic production, additive-free feedstuffs, extra-large eggs, distinctively colored eggshells, and even such minor details as fertile eggs. All of these can add extra value.

Organic food is a very big business now and not without its challenges. Organic certification does come with many measures that govern how the bird is fed, housed, tended, and even marketed. Organic producers of any size are subjected to on-farm inspections and scrupulous adherence to a quite rigid code of production.

This can add most substantially to production costs. There is also still some question about just how much freedom of movement an animal must have and the full parameters of what defines "organic." The organic producer must also be careful about his or her poultry health care and parasite control products. Ill birds are probably just best plucked from the flock as quickly as detected and either put down or totally removed from the farm.

Organic feedstuffs can also be priced at two to three times the going price for conventional feed items. Their scarcity can also result in substantial shipping costs to get them delivered into some areas. Organic production often succeeds best when there is ready access to an informed market segment that is willing to pay a little more for an organic product. This generally means access to a population of at least fifty to one hundred thousand within twenty-five miles of the home farm. The more informed and educated consumers are the better. Organic food items will be for them one more part of a very deliberate lifestyle choice. In rural areas, organic production will sell best around a college campus that is on the margins of urban sprawl.

Many have heard about the four-dollar per dozen organic eggs, but eggs fetching these prices are generally only at large, organic chain stores in high population areas, but demand is growing. Wal-Mart is now selling organic eggs.

With organic production producers must know to the very last cent just how much it is costing to produce a dozen such eggs. To that figure must then be added a fair return for labor and the costs of marketing to provide a return worthy of all your good efforts. The organic egg is viewed by some to be a boutique item. In addition, some consumers have expressed a desire to pay no more than a ten to fifteen percent premium for organic production.

These two points aren't going to be easily reconciled. The organic producer must find a market niche appreciative enough of his or her product to pay the price needed for such production. It will take time to become established as an organic producer and it will require an one hundred percent commitment from the producer and farm.

The use of an heirloom breed to produce organic eggs would certainly add to their appeal. It might even raise their selling price by at least a few more cents per dozen. The consideration with them must be the costs of feed to produce a dozen such eggs. It would seem that the economics of organic egg production at this time would favor only the most productive of the laying breeds. Perhaps where the heirloom layers would really shine would be in the production to produce a "natural" or "classic country" egg.

I realize that the term natural is a real anathema to some who question what does it actually mean? It can mean what you make of it, or more importantly, what exactly you put into an egg carton so labeled.

MARKETING A NATURAL EGG

Simply put, a "natural" egg can mean many things to farmers and the consuming public. Consider using as many marketing opportunities as you can.

Natural may mean additive-free eggs from birds that have been fed no antibiotics or eggs produced only after proper withdrawal instructions have been followed. Read all feed tabs and health care packaging closely and follow those directions to the letter.

Some poultry antibiotic products are not meant to be fed to laying hens at all, but may be used with birds producing hatching eggs. With rare birds flocks, producers cannot completely rule out the use of an antibiotic product, as it may become a producers last line of defense to preserve them in a viable manner.

Poultry farmers may offer what have come to be called "loose-housed" eggs under this heading. Technically such birds may be in a house twenty-four hours a day just as long as they are not caged. They will generally have two to four feet of floor area per bird and may or may not be given range or lot access. The birds are at least able to move about and turn freely.

A range egg should be a natural egg. But just what does "range" mean? It would be expected that range birds are allowed to range over some sort of grassed area for at least a few hours each day. Many will not release their hens from the laying house each day until 10:00 am or later to assure no stolen out nests and lost eggs.

Eggs from hens in chicken tractors are said to be range eggs, but some of those units are as tightly packed as hens in a cage unit. Fifty to one

POULTRY PRODUCER'S SURVEY

The River Hills Poultry Alliance poultry producer survey was conducted over the spring, summer, and fall of 2008. It was circulated at a number of poultry-theme events and farm shows that drew attendees from across the nation. 125 respondents were surveyed with poultry operations in size groups ranging from 46,000 hens to ten or fewer birds. From early on in the polling two groups were discernible; 1) those keeping birds in small numbers as a subsistence practice or for pets and 2) those that are operating or developing a poultry flock as a moneymaking venture for the farm or small holding. Some in the latter group are beginning with small flocks as information sources and supporting infrastructure are still somewhat behind the producer curve in this area.

The first survey question simply asked respondents to assign a numerical value to their level of interest in a number of different breeds and varieties. Some of the bird rankings noted reflect a fairly high level of breed knowledge on the part of the respondents.

The breeds are ranked by the total number of points received (5 being the most interested, 1 being the least interested). Also included is the percentage of forms on which each was cited.

Here then, are the breed rankings: (total points received, overall percentage)

1) Buff Orpington 208 pts		67%
2) Barred Plymouth Rock 208 pts		65%
3) Traditional Rhode Island Red 166 pts		58%
4) Australorp 126 pts		51%
5) Colored Wyandotte 124 pts		51%
6) Dominique 123 pts		48%
7) Delaware 120 pts		49%
8) Welsummer 109 pts		42%
9) New Hampshire Red 108 pts		47%
10) Maran 105 pts		42%
11) Ancona 102 pts		45%
12) Brahma 101 pts		45%
12 a) Brown Leghorn 101 pts		45%
12 b) Cochin 101 pts		45%
13) White Leghorn 95 pts		51%
14) Black Jersey Giant 93 pts		40%
15) White Wyandotte 91 pts		45%
16) Colored Plymouth Rock 86 pts		38%
17) Colored Leghorn 84 pts		44%
18) White Plymouth Rock 83 pts		40%
19) Rhode Island White 82 pts		41%
20 a) Black Minorca 72 pts		39%
20 b) Hamburg 72 pts		39%
21) Other Minorca 66 pts		36%

22) Java 65 pts   36%

The following breeds were added by respondents;
    23) Polish 6 pts   2%
    24) Houdan 5 pts   1%

Red Sex-Link, Silkie, Sumatra, Chantecler, Buckeye, Naked Neck, and White Dorking (less than 1%).

These numbers reflect a well-known bias against white-feathered birds among small and range producers. It is believed that this is caused by many factors ranging from a great many who want just a few birds with a rather exotic appearance to greater predation with white birds on range to bad experiences with white-feathered broiler varieties. Breeds that lay white-shelled eggs do seem to fare pretty well in this poll, however.

2) When asked what poultry varieties were kept, a great many reported keeping two or more different species. The most common combinations were large fowl chickens and turkeys or large fowl chickens and guineas.

large fowl chickens	84%
turkeys	38%
guineas	36%
bantams	34%
ducks	27%
geese	22%
pigeons	20%
quail	7%
peafowl	6%
pheasants	1%

How many species do they keep?

1 species (primarily large fowl chickens)	21%
2 species	22%
3 species	21%
4 species	15%
5 species	2%
6 species	5%
7 species	4%
8 species	1%

3) When asked the total number of birds kept respondents replied:

fewer than 10	9%
10-25	20%
25-50	22%
50-75	12%
75-100	2%
100-200	14%
more than 200 birds	28%

Of total respondents 53% fewer than 50 birds, 14% 50-100 birds, and 28% more than 100 birds.

4) When asked about prices to pay for hatchlings the answers reflected a fairly well informed group.

Baby Chicks		Turkey Poults		Ducklings	
$2-$3	36%	$4-$5	24%	$3-$4	15%
$3-$4	34%	$5-$7	19%	$4-$5	13%
$4-$5	6%	$7-$9	7%	$5-$7	5%
$5-$6	4%	$9-$12	4%	$7-$9	1%

Goslings		Guinea Keets	
$6-$8	19%	$3-$4	21%
$8-$10	11%	$4-$6	14%
$10-$12	2%	$6-$8	4%

We debated including the low figures in each group due to the human nature of some to adopt the lowest price whenever possible, but as a baseline they do show the good number of producers that are willing to pay for quality and rare breed varieties.

5) When asked what to pay for a dozen hatching eggs, ranges were:

$6-$9	32%
$9-$12	13%
$12-$15	14%
$15-$18	5%
$18-$22	8%

6) When asked what to pay for ready-to-lay pullets, ranges were:

$10-$12	58%
$12-$15	16%
$15-$18	5%
$20-$25	1%

7) Here respondents were asked the number of hens they needed to keep for their laying flock needs. The responses were:

Less than 20	39%
20-40	18%
40-60	9%
60-75	4%

75-100	4%
100-150	2%
150-250	6%
more than 250	5%

These numbers are quite reflective of the two distinct poultry owning groups we noted earlier.

8) Respondents were asked where they bought baby chicks (multiple responses were allowed).

Shipping hatchery	44%
Farm supply store	38%
Farmers' market	33%
Local hatchery	21%
Breeders	16%
Local elevator	9%
Other	9%

Lots here with which to conjecture, but many seem to favor local points of sale where they can see what they are buying and can buy in small quantities. A substantial number were making the effort to go right to the source and buy directly from breeders whether on the farm or at an event like a farmers' market.

9) Producers were next asked if they kept any of the following:

Sex-link hybrids	26%
Cornish X broilers	24%
Production Reds	12%
Leghorn hybrids	12%

10) Producers were asked what they sold and they responded:

Brown eggs	72%
White eggs	21%
Breeding stock	18%
Free range broilers	14%
Broilers	5%

Note: Some sold both white and brown-shelled eggs.

11) When asked to describe their poultry operations respondents answered: Note more than one answer was allowed. Multiple systems were in use on many farms and smallholdings.

Free range layers	47%
Day range layers	25%
Poultry yard	25%
Additive free rations	22%
Cage free layers	22%
Free range broilers	11%
Egg mobile	9%
Organic	8%
Line bred flocks	5%
Colony house	2%

12) When asked where they marketed their production the replies were:

From the farm	51%
Farmers' market	32%
Deliver to customers	29%
Restaurants	5%
Through a marketing group	4%

Other (online sales, health
food stores, etc.)   16%

13) When asked the preferred lot size for buying
hatchlings they reported:

10-15	20%
25	28%
50	9%
75	0%
100	2%
over 100	5%

14) This question asked the respondents to rank their
comfort level with the widespread practice of
suppliers offering hatchlings as unsexed or "as-
hatched." A 5 indicated quite comfortable and a 1
as uncomfortable with the practice.

The average of all respondents was 3.3. 23% gave this
practice a 5 while 11% give it a 1.

*For more details about the results of this survey contact Kelly
Klober, 136 Zumwalt Rd, Silex, MO 63377.*

hundred hens in a ten-foot by ten-foot unit are as tightly packed as any other caged birds.

Humanely reared is another one of those nebulous terms. At its simplest it describes birds in a low stress environment with natural ventilation and freedom to walk about, extend their wings, and receive stimulus from their environment.

There are a number of factors that can be packed into each dozen of eggs if the producer so desires and that the market will reward. Farm fresh, additive-free, and cageless are all inexpensively doable and should satisfy the wishes of all but the most discerning of consumers.

Poultry producers will never break through to every group of consumers nor should they try. Many cost driven buyers will always opt for low dollar pricing and represent a market that the box stores solidly have. Why bother slugging it out with such outlets for just a penny or two in profit? The retail trade quite often uses medium sized eggs as a loss leader just to get consumers through the door!

A local group here is now exploring marketing options for some of the eggs from our heritage breeding flocks. We are producing eggs with hues from the chocolate bar brown of the Penedesenca to chalk white Leghorn eggs to the blues and greens of the Ameraucana. Our plan is to package them in the clear plastic cartons just on the market and make the most of all their variable colors. They should catch the eye and we hope to further this with a name such as "Farmers' Choice Hen House Gems."

Will we get it off the ground as we envision? Only time will tell, but it is one way to further build on the rich diversity these birds represent.

From state to state the legal nuts and bolts regarding the direct marketing of eggs from the family farm may vary. The fewest restrictions should be with eggs direct marketed from the farmyard. The more producers reach beyond this zone to market them, especially if marketing them for resale by others, the more you can expect to encounter in the way of rules and regulations.

When probing just what the law does or doesn't say, our local group has found it best to let a third party such as an Extension Agent make the contacts and explore the regulations. The simplest approach for the independent producer with modest numbers is to sell his or her eggs directly to a known and carefully selected group of consumers. In this day and age not enough can be made of knowing your buyers well and culling the disagreeable ones just as you would poor performing birds.

Such a producer may still have to register with the state Department Of Agriculture, receive an inexpensive egg-marketing license, and be subject to on-farm inspection of egg handling and storage facilities. For most of the small producers in our area, the storage unit is a used refrigerator set up in a clean area and specifically dedicated to egg storage. Don't store your cattle vaccines and eggs for sale in the same icebox.

The on-farm inspection normally takes a relatively short time and the inspector's primary focus is on basic cleanliness and a refrigerator unit that operates at between forty and forty-two degrees F.

Such licenses are generally a must if selling to restaurants and retail establishments such as specialty food stores. They may also then be required to have a separate egg retailer's license of their own. Those selling eggs along a classic egg route in a nearby town or suburb may be required to apply for a vendor or peddler's license. Information about these is generally available at the local city hall.

My grandparents sold two hundred forty to three hundred dozen eggs each week along an egg route in the old family neighborhood in St. Louis County, Missouri. They would take the eggs from the flats in thirty dozen-egg cases and place them in brightly colored egg baskets holding six to eight dozen eggs each.

The baskets are either green or yellow in color and display the eggs within to great advantage. At each door they would transfer the newly purchased eggs to the homeowner's various containers. Customers saw the big white eggs between the green or yellow bars of the baskets and just had to have some for their very own. My grandparents were real country folk, selling farm fresh eggs that would become all things good — from Sunday morning omelets to deviled eggs for picnics to Christmas cookies. Who could pass up that combination?

Most of us with modest numbers of birds are guilty of selling our extra eggs in whatever one dozen cartons we have at hand. Into a box will go pink, white, and yellow foam cartons and gray pulp cartons with all sorts of different printing on them. It certainly doesn't look very professional or appealing and regarding the letter of the law, it is not truly legal.

Recycling is generally a good thing, but not with egg cartons. There can be some health and sanitation issues and the lettering on the used cartons seldom accurately describe your eggs. Our eggs are often too large for medium and even large grade cartons and may not even snap closed.

Certainly no thought of fraud is intended, but what is a consumer to make of ten dozen eggs in cartons from ten different original suppliers?

Or brown eggs in cartons marked for white eggs? Or cartons marked for a special class of egg? New cartons will run eight to twenty-five cents each depending upon the amount in which they are bought. The clear plastic cartons are a fair bit higher and are not available from as many sources.

It is possible to buy pre-gummed stickers from various print suppliers that can be affixed to the tops of egg cartons. On to these can be place the farm name, description of the eggs, a brief description of how they were produced, their size, and any other pertinent selling points. In some instances you may have to provide contact information on such a label. By law egg cartons must now also carry detailed information about how best to store and use them to maintain their healthful nature. Pre-gummed stickers for this purpose are also available.

The egg is a most important food item and it just makes good sense to spend the money needed to pack and distribute it in the best possible light. The clear plastic containers would seem to lend themselves to a well thought out plan of sanitation and re-utilization. They also come in four and six egg sizes and unique configurations that could lead to increased small lot sales for today's aging population and smaller families. In setting the price for your eggs don't forget to factor in the costs of cartons, egg cleaning, handling, and marketing. These all add value to your heirloom breed eggs, too.

As noted above producers will need a storage refrigerator for a table egg venture, but that is not all. In that refrigerator should be an easily read thermometer to help monitor its performance. Producers may or may not be required to keep a sheet on the refrigerator door on which to record a daily internal temperature reading for the unit.

As an element of quality control an egg candler will be most valuable, also. I have seen bare bones candlers made from a light bulb fitted inside a tin can a bit larger than the element holding the bulb. A half-inch or so hole is drilled in the closed end of the can over which to then position the egg. Commercially available units range from a small, handheld unit for about twenty dollars to table mounted units selling for several hundred dollars.

The more eggs you have to candle at one time the larger the unit producers need and can economically justify. A darkened room is always best for the candling task and with any number of eggs at all to work through, a comfortable workstation should be created at an elevated bench of waist height or a bit higher.

Candling charts, guides to what to look for when turning the eggs in that directed beam of light were once widely available. One such guide was among the USDA products our Congressman once offered two or three times each year. Old poultry teaching texts such as the classic book by Morley A. Jull carried them also. Contact your local Extension Agent to see what in the way of a candling chart is available through your state university. A wall-mounted chart would prove especially helpful.

We live in a culture in which much of the general population is now far removed from any ties to the land. Some people may be as many as ten generations removed from having any family member working on the land. These people are accustomed to and expecting bloodless meat arriving on pastel colored styrofoam trays and white eggs spotless and stark as bone.

The world of the chicken house is far from a spotless place and your eggs may need anything from a simple wiping with a soft cloth to a full immersion bath to ready them for sale. With only a few eggs being sold to folks you know, a quick dip in a basin of warm water and a wipe down may be all that is needed. With a larger customer base it is time to invest in a bubbler type washer. These are a pail that holds a basket of six to eight dozen eggs and to which water with an egg washing detergent is added. A hose from a small air compressor is attached and the soapy water bubbles around the eggs for several minutes. Figure on paying two to four hundred dollars for such a unit.

A speck of dirt or even dried fecal material will not compromise an egg's internal qualities, but it may be just enough to put off today's often naive and inexperienced consumers. Some are already going to be a bit uncertain facing eggs that are speckled, green or cocoa brown when all their previous eggs have been white. Clean almost to an extreme is now going to be a key egg selling point.

Egg sizes are an interesting subject of discussion for heritage producers. They are set by weight per dozen. The industrialized sector has been built around medium- and large-sized eggs that are the simplest to handle mechanically and package. They are also the egg sizes most commonly produced by today's cage layer strains.

It is said that this is the "super-sized" era and now the big brown egg is a real plus to most consumers. Some customers see it as a "buy" of sorts and those that must limit their egg intake want an egg that really counts.

Some producers have toyed with the idea of producing and marketing some sort of mega-egg. They feed flax meal and some types of fish product to boost omega-3 levels. However, don't say this unless you have the testing to prove it. Other producers have pushed for a truly jumbo-sized egg, but these are not without their consequences. For one, jumbo eggs will require larger, higher priced cartons. They can also take a toll on the birds producing them. A short time back some Delaware lines were selectively bred to produce large eggs and young pullets were lost to impaction and prolapse. There were fertility problems, and it takes a bigger bird to produce a bigger egg, including more feed and a larger house. You must also accept that a flock will produce eggs of several different sizes throughout the course of the laying year.

Egg grading equipment, alas now is made primarily for high volume production. A few years ago at a community auction I saw one of the simple, gravity fed, table top models of yore sell for just a couple of dollars. I was riding with a friend and had no way to get it home unless one of us was left behind and it was his car. Had I known then just how rare those things had become I'm pretty sure he would have found his way home sooner or later.

Small graders now resemble old postal scales and accommodate just one egg at a time. The egg fits into a small cup or box and moves an arm that then points to the egg's particular grade. An alternative solution might be to acquire a weight chart for table eggs and then weigh your eggs out by the dozen on a small digital scale over which it is legal to transact business. Such scales may have to be registered with your state's Department of Agriculture and then regularly checked for accuracy. An egg inspector should be able to help with this.

In trying to build an ongoing market for any farm product, eggs included, a real important point is to supply consumers with information that will put production in a positive light and fully answer the questions of consumers. With eggs this can be done with carton labeling and simple handouts.

This is where the heritage bird producer has a very high trump card to play. Everything from farm name to a simple logo to a simple background sketch of the birds can do so very much to make your eggs a real "feel good" purchase.

Try simple tag lines like "Preservation Acres," "Red Rooster Ridge," "Wyandotte Eggs From One of America's First Breeds," "America's Heritage, the Country Breakfast and the Dominique Breed." Use your words and your breed or breeds to make your point. Inform the consumer that with their purchase they are not just getting a classic American staple, but are helping to preserve historically important breeds and the family farm way of life.

On the subject of meat bird marketing, the temptation is to always begin by bashing the competition. The Cornish X broiler is certainly a big enough target. However, resist the temptation to market in this way. The American consumer seems to want to be approached on such matters from a position of advocacy and not animosity. There are positives enough with the heritage breeds. These were grandma's chickens and that image is rich with positive connotations. Everything from improved muscle tone, sunshine, fresh air, range life, and the odd grasshopper in their diet combine to create a different and certainly more desirable poultry meat.

The task is to get such birds into as many hands as possible and to do so you must work around or through the factors that now influence modern patterns of consumption. The facts are that most of us have limited food storage in our homes now, we are not well prepared to do any sort of processing, we buy and eat a lot of food away from home, and when buying food such as a meat bird we generally do so just one or two at a time.

As a result you are going to have to work every bit as hard to grow a market as they did to grow the birds themselves. The heritage breed birds are going to look different, be offered differently, cook differently and even taste differently than what most consumers have become accustomed to in the way of frying chickens.

Most consumers are now used to buying frying chickens already in serving sized pieces. Often even buying just specific pieces of those chickens. With a dressed range broiler most everything is sold still attached except for the giblets. A great many potential broiler buyers now are going to need something in the way of a book of instructions to even begin to know what to do with a whole broiler. If producers are looking for something in the way of a potential field day to draw range broiler buyers to the farm — how to cut up and freeze a broiler might be one good demonstration topic.

How can producers best market a farm fresh broiler, in this age of nuggets, strips, and drive-thru windows?

Marketing a Broiler Chicken

The following are some tips to help form a broiler-marketing plan.

• Start small. That means both in numbers and expectations of early sales.

• To a great many consumers, broiler producers are offering something that is very new and very different. Expect many early sales to be of no more than one or two birds at a time. Realize that not everyone is a candidate for such purchases. Even as a customer list becomes established, seldom will producers encounter buyers taking more than five or ten broilers at a time. Today's families are smaller and the freezer space usually sits atop the family refrigerator and is measured in a very few cubic feet.

• Producers and potential customers need to be on the same page as to what a "broiler" really is and what they want to see bred and fed into it.

Every time Tyson or KFC takes a hit on TV during a sweeps month, interest will go up in an alternative meat bird. Every time the words "avian flu" are mentioned in the news the winds blow cold on producers and their birds.

• To understand the pricing behind the bird customers must understand the bird and what went into it.

The $2.50 a pound range broiler draws a lot of small farmer interest, but is that price truly sustainable? Can it also be sustained across a broad spectrum of the buying public? The natural broiler is a premium product, but not everyone wants or can afford such things. Grow out only the number of birds that can be marketed at the fullest possible profit potential for your area and market them no more.

• With the heirloom broiler there will be a teaching job and possibly a quite big one for consumers. Questions from "Why doesn't it look like those birds in the store?" to "What am I supposed to do with it now?" will have to be answered.

• Take a cue from some of those high dollar catalog companies selling braces of pheasants and smoked ducklings. The broiler you have to sell will be just about as alien to most consumers and just as rife with expectations. Consumers believe it to be something quite good, but will need help to achieve all of the expectations that they have for it. This educational task will best be done in multiple parts and can include signage, business cards, flyers, handouts, instructional pamphlets and more. The simplest place to begin may be with a basic handout on how to disjoint a bird. Also include freezing instructions and a couple of easy recipes.

- Regarding advertising, the two big questions to be answered are how and where? Answer those and you will pretty much have answered a third, how much?

- Sadly, we have come to a time when you will want to be very careful about how much attention you call to yourselves and whose attention do you try to attract. Begin ever so slowly to release the word about the venture and try for as much gratis coverage as possible.

- Send a short and simple press release about the new venture to your local, weekly newspaper. They are always looking for positive locally slanted material with which to fill their pages. A well-written press release may draw someone out to do a feature article and will certainly help to build word-of-mouth.

- Be your own best salesman and promoter. Take new and novel chicken dishes to church suppers, fundraisers and civic events. Hand out business cards freely and expect a bit of quid pro quo from those with whom you do business. If I drive your tractor you should be eating my chicken.

- You often hear the expression "priced to sell," but what should really matter is "priced to profit." When my grandparents owned their little confectionary, the standard markup on food items was thirty percent. That might not be too bad a guide as to how to price dressed broilers for the long term. A thirty percent markup over total costs to produce (including labor and all overhead) should create at least the base point for pricing.

Factored into the costs to produce a range broiler must be the area set up for processing, slaughtering equipment, scalding apparatus, and a mechanical picker. Even with accelerated depreciation their maintenance and use will add to the costs to produce year in and year out.

The broiler, roaster, and heavy baking hen markets are all going to be somewhat high labor ventures serving rather niche markets. Processing and marketing alone may take up a full sixty percent or more of the man-hours allotted to them. Achieve too high of a profile and level of production and independent range broiler producers will also draw the high volume, industrialized producers to this market arena.

I have talked to producers who grew their broiler ventures to the five thousand to ten thousand birds per year range and they were almost unanimous in their belief that it too quickly became too much for their smallholdings and themselves. From the disposal of feathers and viscera to coping with a broader, and ever less well known to them, customer base — the problems grew right along with the bird numbers.

It seems to me that the happiest and the most successful producers are coming through at the level of five hundred to two thousand birds per year. For them broilers remain but a single venture in their enterprise mix and are not a business entity bidding to dominate the whole farm. In part the heirloom producer will be endeavoring to market surplus males and older culled pullets as broilers and roasters. So these producers may even have numbers of fewer than five hundred per year to market.

Hopefully, research will be launched soon that will substantiate the anecdotal evidence that has shown that some heirloom breeds meat birds have quite distinctive taste and texture differences from the modern commercial broiler that can be accessed through local supermarkets. Imagine what could come from the Dominique or other American heritage breed being able to legitimately claim that it is comparable to pheasant on the plate?

HATCHING EGGS

Spend even a short time online now and you will see just how big a business the hatching egg has become. Part of it has been because of the refusal of some online marketing services to accept live animals and birds, but the greater part has been because of the marketing opportunities such eggs have come to again represent.

Should a breed catch fire, the first thing most producers of that breed have to offer for sale is hatching eggs. Along with live adult birds they are also the marketing option with the longest history. A few Septembers ago all live poultry shipments were removed from the U.S. Mail for several weeks and many still fear that this could yet become a permanent thing. The airlines seem almost eager to end such shipments and the animal rightists people have come to view this as one more potential victory for their side, too.

Thus the hatching egg has history on its side and one of the more assured futures as a conveyance of poultry genetics over wide distances. Those that object will have a bit of a challenge making a case for "egg abuse" in a society that enjoys eggs scrambled and over easy.

A great many individual breeders also seem more comfortable marketing and shipping hatching eggs than baby chicks, started birds or adult breeders. Most have had experience shipping or receiving small lots of hatching eggs. There are even sources now offering to custom broker hatching eggs and help with shipments.

I have seen hatching eggs sold in quite small lots down to just two or three of some of the very rarest of breeds. They may even be offered in a mix of hatching eggs of several different breeds. These kinds of small lot sales certainly have to appeal to those working with very small numbers and who are unable to ship even minimal lots of fifteen chicks with any consistency. It can also develop some of that always important, early return on investment. Thus far all post offices will accept hatching egg shipments at any season of the year.

This market receives some definition from the online hatching egg auctions. I have seen reports of the eggs of some of the rarest of the rare bringing in as much as ten dollars each, but most heirloom eggs bring six to twenty dollars per dozen plus a packing and shipping fee of six to nine dollars per dozen.

Hatching egg auctions can be a very expensive way to acquire a very few chicks, but it has often been the only way to access certain breeds or access blood from flocks a great distance away. Most old hands have tales of scrimping and saving and waiting months if not years to acquire even a handful of birds from one of the name flocks half the country away. Such flocks are far fewer in number now, but the hatching egg trade is still very much alive.

Most pursuing this particular marketing option have developed their own special touches for use in the always-tricky shipping process. I read about one breeder's method that begins with acquiring the segmented cases used to transport whiskey and other bottled spirits. These are strong, well made cartons reflective of their expensive contents, but still a handy size for shipping eggs in the mail.

Each segmented area can be padded heavily and a single egg centered midway in the box. They can be padded to accommodate two eggs per segment. Thus such a box can safely accommodate one to two dozen eggs per shipment with heavy padding for the eggs.

Sellers certainly cannot guarantee the hatchability of a shipped egg, but it is a fairly standard practice to offer to replace eggs broken in shipment. Such losses are generally not the producer's fault, but it is a gracious thing to do and should help to build sales.

Better packaging methods for shipped eggs are coming and I recently saw a neat lined shipping box with egg-sized recesses in the foam. Cost will certainly be a factor with such products, but as shipping containers improve, the hatching egg market should grow stronger as a result.

The hatching egg market is quite seasonal at the moment. Prices and demand changes as the year progresses. The demand for certain breeds also declines as they make their way into more and more hands. I feel that this market still has much upside potential, but in the future prices will be based on the quality of the birds behind the eggs and not because of the novelty of the breed.

For the whole of my life, the baby chick has been big business, but this did not occur until the development of dependable, larger capacity incubators. In the early days of artificial incubation most chick sales were local and direct.

The ability to send chicks through the U.S. Mail was what really spawned the poultry trade, as we now know it. Artificial incubation, improved shipping methods, and the emergence of the White Leghorn all came together early in the twentieth century to begin creating what was for decades, a farm based industry. Cities like Petaluma, California, became real hubs of poultry farming activity.

With the mail rapidly moving along baby chick shipments, now breeders could share rare and valuable genetic pieces from coast-to-coast. Hatcheries could reach beyond the borders of local townships to market their birds. I can still recall the last two local hatcheries in our area and the economic force that they were in their communities at one time. They both operated with modest breed lists heavy on the traditional Red and Rock breeds.

Following the absorbing of the egg yolk shortly before hatching, the baby chick is good to go for up to ninety-six hours without feed and water. That means that shipping to just about anywhere in the lower forty-eight should be doable from just about anywhere else in the lower forty-eight. Alas, this is not the case for states such as Montana where access to larger air traffic hubs is not possible. Watch for temperature considerations with baby chick shipments even though the little chick seems to be almost made for shipment across rather substantial distances.

Chick shipping boxes with their absorbent bottom pads (sold separately) are now available from several sources in lots as small as five or ten. A one segment, twenty-five chick box with pad can run as much as two to three dollars. This box though does much to keep chicks warm and secure in transit. New hydrating gel products have been developed that can be placed in the shipping boxes for the chicks to pick at while in transit. Such a product can keep the chicks hydrated and perhaps even more content while in the mail.

Huge numbers of hatching eggs and baby chicks are being marketed on the Internet now and the appeal is that you can create your own marketing campaign there. However, take much of what you see and read there with a grain of salt. There is advertising and promotion, there is hype, and then there is pure fiction. Try not to be guilty of either of the latter two.

All along my concern about the Internet has been that it is addressing people with all different levels of experience and expertise. Just because someone has a website or loud voice in chat rooms doesn't award master breeder status. There are some birds online now that are very dubious and some online that can be accessed in perhaps no other way. I believe how these are finally reconciled will ultimately tell how successful and enduring this marketing option will be.

Please, please be honest with any advertising and promotional efforts. When birds are presented either for sale or exhibition they are representing the whole of their breed and all the folks who are working with them. If producers want to add value to them, then provide as detailed a background of the birds as you can to all potential buyers. If they need some work, say so. If they are not yet one hundred percent pure in their breeding, say so. If their level of productivity is not yet what you wish it to be, say so.

At a farm show a couple of years back, I set up several trios to sell and displayed them with cage front flyers recounting their strengths and their weaknesses. Many took the time to read the flyers and then study the birds even more closely. All birds were sold, sold at quite good prices, and I drew many favorable comments for presenting the birds honestly. I have seen a lot of angry people and hurt feelings from being sent birds that weren't as represented. I have to wonder how many badly needed new breeders have been lost forever because of such practices. Too many times I have been asked by beginners to go through their recent purchases and then had to tell them that they have been sold junk or even the wrong breeds of birds.

With bird promotion it is always best to be just as honest as possible. Explain coloring variables, fully addressing shipping matters and costs, advise if supplies are limited, and describe the birds for sale as they really are. If buyers are realistic in their expectations at all they are not expecting national show winners nor catalog picture exemplars. They should be very much true to the breed and of the same quality as you would keep for yourself. However, we all have our share of the "less than the best" and those are what birds swaps, sale barns, and chicken barbecues are for.

At one time even the most broad based of farm magazines were full of ads for individual breeders selling chicks and eggs from their performance bred poultry flocks. There were even national magazines built around a single breed such as the Ancona.

A few breed clubs still do publish their own newsletters and yearbooks and there are again a growing number of poultry themed magazines and newspapers. Established periodicals such as *Acres U.S.A.* are also giving over more space to cover poultry themed material. A breeder's listing in any one of these would certainly give your heirloom flock greater exposure.

Direct sales of baby chicks at events such as farmers' markets are a close to home marketing opportunity that should not be overlooked. At our market we sell chicks from early spring through mid-fall, but the strongest markets are in spring and early summer. These kinds of direct marketing opportunities seem to work well for both buyers and sellers.

There are a lot of country folk who want to see what they are buying and then to buy them in the exact numbers that they want. I will sell a group as small as three or four chicks if they are going to be taken straight away to a good home. With such small lot sales I also expect to be paid quite well for them as the takeaway chicks are sparing the buyer shipping and set volume purchase fees. Another plus for such buyers is the opportunity to talk one-to-one with the producers about their questions and further bird needs.

MARKETING BEYOND THE LOCAL AREA

If you are going to market beyond the local level then there are more issues to consider. The greatest demand will be for chicks hatched fairly early in the year. Some customers will want January hatched chicks to be sure that the birds produced from them are ready for summer or fall shows or autumn laying.

Will your particular facilities enable you to be in production that early in the year? Make it very clear to all concerned when chicks will and won't be available.

Will the expected numbers be available? Most sales will be for fifteen or twenty-five chick lots. State up front if only very limited numbers will be available. If there is a waiting list, make sure everyone knows that it is there and how it operates.

Most now charge for postage over and above the selling price of the chicks. Check with a local post office about their acceptance of live shipments and the cost to mail a box of chicks to different areas. It will probably be best to ship heirloom chicks via priority mail.

Do not fail to include box costs in the chick pricing.

Producers can hatch to order, but figure on setting something like thirty percent more eggs than chicks you will need to fill the order. This may leave producers with surplus chicks with which to deal with, but it is a time-honored practice to always add one or two more chicks to the box than is called for by the order. Top that figure by another couple and watch the buyer goodwill build.

If a hatch fails, get on the phone and be upfront with those expecting the chicks. Most will be understanding and accept a later shipment, but do offer to return any advance payments they have made. Many expect a third to a half of the selling price as an advance payment on all orders hatched to order.

Sort the chicks very closely before placing them in the box for shipment. To build the business it is important to send out chicks that are sound and vigorous, of good size, and are of good type.

Hatch out on a Sunday or Monday to prevent the risk of chicks getting hung up in transit somewhere over a weekend. Also be mindful of when holidays fall during the hatching season.

Pricing can be a tough determination for some folks. Most farm people are modest sorts and easy going (if not perhaps too easy going sometimes). If a hatchery charges $2.50 for a good Buff Rock or $4.50 for a Welsummer chick then one from a fellow farmer should be worth no less. The locally produced chick could very well be worth more if that producer has been selectively breeding for improved performance. Producers can take the price as a guide from the commercial hatcheries, but don't be limited by them. Most independent breeders price their chicks in the $3.50 to five dollar range.

I have set up at swaps with two or three very rare breeds not even noted in the leading catalogs. Greenhorns will come by with a copy of a catalog under one arm or tucked into a hip pocket. When they cannot find our birds in those catalog pages they are either at a complete loss or in a state of disbelief. The poultry world is a broad one and not even the largest of hatcheries can expect to catalog and offer every breed and variety.

As the year progresses chick and egg prices do tend to go downward and there comes a point when it is better to turn off the incubator than damage your market with over production. Success may even spark local imitators and competitors, some doing it with birds bought from you. Such is the nature of capitalism. When I am told someone is offering birds similar to mine for less money I have a stock answer, "Who knows better what they're worth than the one who produced them?" Remember that one of the things producers set out to accomplish when they took up a rare or endangered breed was to get them into more hands.

Competition really is a good thing if it is a motivating factor to do a better job with birds. The real challenge is to make next year's chick crop better than this year's. If you do that, your position is made secure, your place in the market held, and your potential for increased sales naturally increased.

Started and adult bird sales would seem the ultimate marketing option to which to aspire. I once spoke to a representative of a major hatchery who espoused a long-term goal of his firm to move entirely to the marketing of such birds.

It would certainly pare the costs of incubator operation, but it is far from a lost cost venture. Though, there is something of a history for it.

The "started" or "ready to lay" pullet has been sold for decades to small and medium-sized laying flock owners. These are twenty to twenty-two week old pullets ready to go straight into the laying house. They are sold for several times the baby chick prices and were generally available in all but a handful of breeds — mostly White Leghorn strains and their hybridized crosses.

Buyers could justify the higher per bird out-of-pocket costs by realizing the increased costs for brooding, grow out and maintaining the needed facilities. These birds also have the owner in production in a matter of days or weeks instead of months.

A friend of ours now grows out five hundred to one thousand sex-link pullets each year. He buys them as day-old chicks in lots of two hundred to four hundred (some price discount may be available for purchases at this level) and begins selling them at twelve to fourteen weeks of age. They are generally all gone by twenty weeks of age.

He sells the young pullets for between five and six and a half dollars per bird depending upon their age. You can buy one or all on a first come, first served basis. Good purebred pullets twenty-two to twenty-six weeks of age should bring $1.50 to $2.50 more per bird. We recently sold some Rosecomb Rhode Island Red females for $12.50 each. Even a common brown egg-laying hen going into her second or third year of laying will bring four to six dollars each at most swaps.

The rare and truly better birds really have no price yardstick. Recently we've seen $35.00 and $125.00 trios of different heirloom birds sell quickly when set up before a general farm crowd. There are show breeders now asking for and getting over one hundred dollars per bird.

They're not selling lots of them, but at those prices they really don't have to. To be realistic, however, this will always be a somewhat limited market niche except where you can develop a localized trade for started pullets.

Costs to ship via the mail will generally be as great or greater than the price of the birds. If you can get to where large numbers of potential buyers do gather, they represent a potentially strong market for good trios, breeding pairs, and even single individuals.

Started chicks are generally birds between three and eight weeks of age. By this age most breeds are fairly easily sexed, they are in good feather, costs to finish out will be lower, and the brooding task will be largely behind them. For all of these reasons they will not be cheap and they are still far from made birds. Expect to pay at least double baby chick prices for them.

This is the gawky stage in a chicken's life. They are often not at their most visibly pleasing nor are they often readily available. I offer a few such trios and pens (one male and four females or two males and eight females) for sale each year, but both older and younger birds seem to outsell them. Buyers often try to buy just the pullets from such groups and leave the males.

One of the reasons for offering this particular marketing option is to help dispose of some of the extra males of the year at a fair price. When we set up trios and pens for sale we price them on a per bird basis for taking all of the birds in the group. When buyers ask for just the females I leave the group price the same and say I'll just give you the rooster. Many times they reply I'll give you that price, but you keep the rooster. That deal I will take! I don't mean it to sound flip, but neither do I intend to wind up with dozens of unneeded young cockerels at the end of the year. If the rooster comes out — the price still stays the same. My daddy didn't raise any slow children!

The market for heirloom fowl now has many similarities to the so-called exotic bird and animal market of ten to fifteen years ago. There is always a breed or color of the moment, prices run up greatly in breeder markets then tumble as numbers build. More breeders committed to the long term are needed, and there is still a need for more fact and less hype in breed promotion.

Still, it is that energy and enthusiasm that has, at least in part, brought the heritage birds to this raised point. Whether producers are marketing to foodies, fellow small farmers or rare breed fanciers, there has to be this energy, this passion for the birds and their future.

If this plan of preservation and heritage breed/family farm promotion is carried out successfully ten years from now we should still be selling three to four dollar baby chicks and even a fair number of fifty-dollar roosters. We shouldn't be breed shopping for the next Dominique or Java nor making a fad of feather colors. Underwriting all heirloom breeds in the future will be their value as producers of eggs and/or poultry meat. They will be quite artisanal in nature and a fine poultry trade will emerge just as what we are seeing today with fine American wines.

Very much a part of their future will be keeping flocks small and managed exactingly. Also important will be independent producers coming together to make a common cause when needed, though most marketing is made directly to end consumers. In addition, birds need to be consistently bred to an elite yet practical standard. Greed is not good in this type of production agriculture. It will kill the hen that lays the golden egg. More precisely, it will kill the market for those golden eggs. As simply as I can state it, your successful marketing plan will hinge on just one thing — knowing when to shut off the incubator.

OVERCOMING CHALLENGES WITH RARE BREEDS

Situated in Barrington, New Hampshire, Yellow House Farm celebrates the rich cultural heritage of farming in New England. Co-owners Joseph Marquette and Robert Gibson are proud to offer the benefits of locally grown, healthy food to their community — in a traditional way just as those in previous generations have done.

Both Joseph Marquette and Robert Gibson grew up homesteading. For Joseph, as a child, birds of many types had always been around: chickens, ducks, geese, guineas, and turkeys. Returning to rural life meant chickens every bit as much as it meant vegetable gardens and sunshine. They were coop building and garden ploughing before they even started home renovating.

Birds had captured Joseph's imagination as a child, although other farm livestock were also part of the picture. As a little boy, he used to ask his father to catch the chickens and throw them into the air so that he could watch them fly. Joseph revealed, "There's something about flight that has always amazed me. I used to have a flock of white muscovies that would soar, drakes and all, about the yard, roosting high in the maples and pines about the house. They were spectacular to watch."

When the decision was made to move back to the country, Joseph, who took charge of the poultry operation, started looking for chickens and found that there were none to be found — anywhere. This sent him into research mode. The more he started looking through books and on the Internet, the more he learned that there were so many fascinating breeds to be had, breeds that had nourished and accompanied man through the ages. Sadly, he also learned that many of them were on the brink of extinction. This seemed at the time, and still does, to be such an utter pity as well as an indictment. Shocked that we are permitting the finest achievements of our collective agricultural history to disappear in an unnoticed swansong of neglect, Joseph searched further and discovered the Society for the Preservation of Poultry Antiquities (SPPA).

He become a SPPA member and gained access to their breeders' directory — which was a real treat. For Joseph, the SPPA has really turned out to be a great community of individuals whose love of bird and lore are a profound source of guidance and inspiration; he says, "We're all in this little ship of preservation together."

The farm was purchased during the summer of 2005, with chicks arriving a week before they moved into the house. Not the best arrangement, but Joseph Marquette's parents were able to play "grandparents" and look after the young chicks for a week, while a coop was hastily constructed. Why the rush to have chickens on the new farm? Well, moving back to the country just meant raising chickens and fundamentally, chickens represented the essence of what it means to live in the country.

Joseph and Robert experimented with many different breeds. Time and again they came to the same realization — few have been working with our dear heritage breeds for the development of their productive farming qualities for quite some time. Universally they found that the usual sources of breed information are but reference guides or histories. They point to where things should be but not to where they are. Joseph says that, "Finding dynamite strains of productive heritage fowl who live up to their supposed past glory is like the proverbial needle in a hay stack."

Recognizing that they would be starting from scratch and being by nature rather tenacious, Robert began researching every old manual he could discover. He found that every breed had its champion and braggart, but as he continued through the poultry manuals he began to see categories of excellence: good foragers, flighty and able to fend for their own, docile and easy to tend, fast growing, early maturing, winter laying, broody, non-broody, excellent meat quality.

On the Yellow House Farm, poultry are divided into egg production and meat production. For egg production, Joseph feels that Mediterranean and Dutch layers are not to be beat. He says, "Barring oil-dependent, mono-generational, entirely unsustainable hybrids, they are the true option for any farmer hoping to earn money through the sale of eggs." On account of the northern New Hampshire climate, Yellow House Farm requires rose-combed or other-than-single-combed birds. They opted for Rosecombed Anconas even though their bloodlines were limited. Hence, they began with rosecombed cocks and single combed hens. These highly separate bloodlines are just what is needed for invigorating the stock; in addition, fertility is very high, and production is impressive. For Yellow House Farm, the Rosecombed Anconas lay at least 33% more than their Dorkings. In 2009 they are beginning to launch the same breeding program with Rosecombed Black Leghorns.

For meat birds, Joseph found that the variety of choice was impressive, but not all were created equal. He found that those that grow the fastest generally do not possess the same quality of flesh. Ultimately, between the two, the latter was the more important to them. Their goal is to produce meat of exceptional quality. If that means relying on birds who reach their best at 20 to 24 weeks, then so be it.

Joseph noted that, historically, five breeds have been praised more than any other in the realm of flesh quality: the Dorking, Houdan, Crevecoeur, La Fleche, and Old English Game. These are not as large as the Jersey Giant nor as fast growing as the New Hampshire, but they are deeply delicious, fine-boned, with fine-grained flesh that is exceptionally juicy. Yellow House Farm customers rave about their flavor.

Of these, he says that the Dorking has come to the fore, although they still maintain a small flock of Houdans. "It is not that the Dorking is of finer quality," he says. "In truth, they are more ready to be the mainstay of a meat program at the current time." Yellow House Farm has found that as they grow as a business, it is ever more difficult to maintain multiple breeding flocks in multiple breeds. Moreover, commercial success often depends on brand stability.

Vigor and general health are the top priority of any heritage farmer. Yellow House Farm birds are expected to thrive out of doors in virtually any weather. Those that do not are culled immediately. Joseph comments that immunodeficiency is the result of the irresponsible raising of fowl indoors where they do not come into contact with nature in such a way as to build up and maintain powerful immune systems.

The general order their selection runs is as follows: vigor, weight/size, conformation/shape, and lastly color. Of course, any bird showing a defect, e.g. cross bill, wry tail, malformed toes, etc., is automatically culled. Joseph maintains that it is much more important for the farmer to retain birds of proper weight than that they be perfectly colored. If one consistently favors color over weight and conformation, one's production will plummet.

In general, Yellow House Farm practices roll-back breeding. They have pens of fathers on daughters, sons on mothers. Each year they reduce their current hens by 50%, saving the best and the brightest, to make room for the incoming pullets.

They are also beginning a system of grading between varieties. By taking a group, for example, of Red Dorking hens and covering them with a Colored Dorking cockerel, they can hatch chicks that are 50/50. By breeding those 50/50's with another Colored Dorking cockerel and then continuing this pattern over multiple generations, they can build up a new line of Colored Dorking cockerels that will benefit from the underlying genetic diversity provided by the original Red Dorking hens.

Yellow House Farm seems to attract challenges as its specialty. Joseph says that in New Hampshire they do not yet have a state meat inspection program. Luckily it is currently underway. This will open up the possibility of further growth and venues of production.

He also states that procuring stock in rare breeds can be rather tricky. The Society for the Preservation of Poultry Antiquities (SPPA) is an essential resource. However, often breeders become very jealous of their birds and maybe a bit impractically sentimental, this to the detriment of the breeds which are always in need of further stewards.

There is also a constant need for funds reinvestment in infrastructure, etc., at Yellow House Farm. Sometimes they feel that they are going to be building for the rest of their lives. Joseph also mentions the difficulty of the multiple farmer's roles as accountant, PR personnel, secretary, market vendor, coop shoveler, stock manager, hatchery runner, butcher, and so forth. The many hats to wear are at times daunting.

However, at Yellow House Farm, Joseph says the greatest obstacle to good, local food is by far the sea of unelected government bureaucrats who get in the way of everything. Local health inspectors can be a dream in one city and a nightmare in another. Looking at the same regulations, they come to different conclusions. Friend or foe, few of them really know what they're talking about with regards to heritage poultry. If they have any experience at all with poultry, it is invariably with factory, i.e. agribusiness, settings that in no way reflect the reality of the heritage poulterer.

On the other hand, Joseph counters, we have often found elected officials to be good listeners. Eventually, we as a community of eaters — producers and consumers — will need to work for regulatory system reform in order to remove the obstacles that separate local farmers from local consumers. Furthermore, land-grant universities must return to their roots and focus on the needs of the surrounding communities so that graduates leave these institutions with a clear view of what is needed to develop clean, sustainable, and local food economies based in traditional methodologies and stock that is compatible with environmental needs and the good of our public health.

When asked about what advice he would share with would-be heritage poultry farmers, Joseph could barely be contained. Here is his list:

1. If eggs are a major part of your farm, raise a heritage egg breed: White-Faced Black Spanish, Anconas, Andalusians, Minorcas, non-commercial Leghorns, even Hamburgs. The White-Faced Black Spanish is a breed that simply must be restored to its egg laying glory. This is a breed that has been feeding us for millennia. It must not be lost to oblivion or one-sided breeding just for fancy points. They would certainly be the Farm's heritage layer were it not for our northern clime. First to arrive with the Conquistadores, they far out-date the Leghorn.

If you choose an egg breed, you will also need to foster a market for your excess cockerels as fryers, which is their rightful place in the kitchen.

2. If meat is in the equation, choose a good meat breed. The best are to be found among what are typically referred to as "general purpose" breeds. It is probably better to avoid larger breeds such as the Jersey Giant, Brahma and Cochin. They are too big for their own good in many respects, and you'll eat your shirt trying to feed them to market size.

The Dorking in all varieties is to be recommended. It is an excellent meat bird with fine table qualities. Other breeds ready for serious consideration would be:

the Faverolles; the Sussex; the Plymouth Rock in the Barred, Buff, and White varieties; the New Hampshire; the Wyandotte, probably the White, Silver Laced, and Columbians are best; the Chantecler. The Orpington in at least the white and buff varieties is another strong breed. The Dominique is probably another good option: a tad smaller yet well formed.

All of these breeds also lay well, although none compete with the Mediterranean varieties. Still, the extra eggs they lay will augment your egg production and sales, and this helps to cut cost drastically.

3. Now that you potentially have two breeds, your plate is pretty full. Most certainly do not fall into the trap of having a big mixed flock of everything: stay with one egg breed, one meat breed. If, however, you can't resist just one more thing, do a good deed. Choose a truly rare breed that traditionally possessed admirable farming qualities and nurse it back to health. Make it powerful. Hold it to true farming standards. Consider the La Fleche, Houdan, Crevecoeur, or Redcap as ancient foundational breeds in need of heroes. If you'd like a rare, modern composite breed, consider the Buckeye; Partridge Plymouth Rock; or Partridge, Black, or Buff Wyandotte. The Old English Game might fall in here to open up a specialty niche on the side for meat of excellent quality in spite of its more petite size.

4. Remember that no heritage breeds are truly up to snuff. We have been neglecting them utterly, and it shows. Nonetheless, it is possible to find specimens in the aforementioned breeds that are ready to be brought back to center stage in a serious way. You're going to work, but that work will also be your pride and a gift to your great-grandchildren.

Any serious heritage poulterer is going to breed his or her own stock each year, following carefully considered selection standards. If you buy your stock from hatcheries yearly, your birds and, thus, your product will never improve. The improvement of your breeds rests in your hands. Once you've chosen the fowl you intend to raise, acquire stock from several different strains in order to infuse your breeding program with vigor. Remember that we're starting over. Acquire some worthy bloodlines, decide what you're breeding for, join the SPPA, and then off you go.

5. Undo what is not real when it comes to ideas about poultry. As it stands, food culture has been turned on its head. Very few modern consumers maintain any of the prejudices of the past that are often touted in poultry books. Moreover, it's better so; most of the past prejudices had nothing to do with anything of value.

First of all, at least in New England, the question of yellow skin or white skin is of no importance whatsoever. All of our chickens are white skinned, and no one has ever mentioned it to us at all, ever. Moreover, considering the demand for our meat, I doubt that they ever shall. Besides, were anyone to ever notice and ask about skin color, it would be an excellent opportunity for education about our breed.

Secondly, egg color is another non-issue. On the rarest of occasions someone will look surprised to see that our eggs are white, yet no one is ever concerned about it. Indeed, once I explain that all of Europe's ancient fowl lay white or slightly tinted eggs, they perk their ears in interest, another great educational opportunity.

Then, there is the question of pinfeathers. Many of our birds are dark feathered, and on occasion some trace of dark pinfeathering will remain, although we remove as many as possible. When asked about them, we explain that the vast majority of heritage fowl are beautifully colored and that modern markets have conditioned us to think that everything is white. By doing so, they have put most of our traditional fowl in jeopardy. Never have we lost a sale because of a few pinfeathers. To give you an idea, our two market ducks are Cayugas (black and blue varieties) and Muscovies (black and chocolate varieties). We sell them dressed by the hundreds, even with a few little black pinfeathers. It should be noted, though, that butchering at the correct time does a lot to reduce the issue. Furthermore, if the occasional bird just doesn't seem neat enough to market, it gets to grace our own table, which is a nice treat.

6. Choose your name carefully, and then use it everywhere. Make it fun and unique yet not too wordy or out of the ordinary. Your name is your first step in branding. You want everyone to remember it and recall it without difficulty. Aim for specific qualities and then maintain those qualities such that your name becomes synonymous with those qualities.

7. Packaging is important. Develop a label that reflects your farm and then stick with it. Often eggs are sold in recycled supermarket cartons with the original names scribbled out or covered up. Poor presentation gives an undesirable first impression. A nice clean package with proper labeling signals to customers that you're a professional. This is an important note; you are a professional farmer, even if you're small-scale. Claim that and then present that.

8. Know your fowl well: their history, their productive qualities. Speak about them whenever you have a chance.

If you're shy, get over it. People are interested in food, especially good food. When they part your company with a deeper understanding of your heritage poultry, they will relate what they have learned to others. It's among the best advertising.

9. Learn to cook your fowl well. Customers will have endless questions. If you cannot answer them with distinct, clear, and simple answers, they will be discouraged from buying your product. Moreover, if you do not know your birds' cookery well and send them down the wrong path, their disappointed tongues will carry bad advertising on their tip.

10. Sell directly to the customer whenever possible. Your profits are eaten up by the middleman, whose chunk will always be disproportionate. Besides, attending farmers' markets allows you to educate your customers about your poultry. These opportunities are invaluable.

11. Know who your customer is. Your product has a value that must be paid. Heritage eggs raised on free-range are simply better than any alternative, organic or otherwise. Heritage meat is so utterly superior in quality to conventional broilers that the difference is evident to any palate.

Your products are simply more costly. Your customer is the one who recognizes this and is willing to pay for that quality. Keeping track of costs and demanding a fair price is not swindling; it's proper business.

At times we see the occasional market attendees mutter about the cost of our product, but that's all right. Either they are new to local food and haven't yet learned all of the ins and outs of what we do to bring them this great food, or they're still attached to the idea that food should be unspeakably cheap. Most will awaken from this over time and with the proper exposure to others who have already learned about real food. Others simply will not. Their refusal is not a reflection of your product's worth, and you cannot feed your neighbors for free.

As it stands, in the current market economy of 2009 here in New England, heritage chicken is worth at least $4.00/lb, and heritage eggs should be sold for no less than $4.00 a dozen. It should be noted that these prices are rock bottom.

12. Love what you're doing! Your customers will sense it and appreciate your product all the more. Moreover, it gives you energy and shores you up for moments of trial.

13. To balance everything isn't easy. We have found that the rhythm of farming life is not the same as that of more mainstream professions. It requires seven day a week attention, three hundred sixty-five days a year, and in truth we put in very long days. The tradeoff lies in the many moments of stark awareness: being there to see the bloom of flowers, the hatching of chicks, the flight of barn swallows, the smells of the earth. I'm actually not trying to wax poetic; that is the blessing.

I save time every morning for quiet and centering. I read a chapter in a book that inspires me, and I sip coffee in stillness. If I let this practice slip, I find that I begin to flail about, becoming ever more burdened by the never-ending list of chores. Centering myself daily helps me to remember the gentle path even when I have to work very hard.

Praise each other and try not to complain. Sleep enough and eat the right foods. If your chickens will stop laying because their food is rubbish, why on earth would you be otherwise? Eat well. Sleep enough. Cultivate gratitude, and desire joy. On occasion, listen to music and open a nice bottle of wine. Get rid of your television. Sit on the porch and watch the fireflies instead.

Joseph Marquette and Robert Gibson can be reached at Yellow House Farm in Barrington, New Hampshire by phone at 603-335-6131 or visit their website at http://yellowhousefarmnh.com.

CHAPTER 9

WORKING TODAY — THINKING ABOUT TOMORROW

BUFF ORPINGTON
• MALE •

BUFF ORPINGTON
• FEMALE •

RHODE ISLAND RED
• MALE •

RHODE ISLAND RED
• FEMALE •

I SAT ONE AFTERNOON A FEW MONTHS AGO WITH A GROUP OF PEOPLE WHO HAVE BEEN ABOUT THIS POULTRY THING AT ONE LEVEL OR ANOTHER FOR DECADES. Most had started as youngsters, growing up on small family farms where chickens had been kept for profit and sustenance.

Chickens had been a part of their lives and mine through good times and bad. We had held onto them when they were supposed to no longer be of value in small numbers on family farms. None of us were making our whole living from the birds and some would even qualify as old time chicken traders. This shows though, that the money never was completely wrung out of the small chicken flock.

As we sat talking at the end of a poultry market in a little city park in Mark Twain country in Northeast Missouri, the question was asked, "Will the chicken thing last this time?" The consensus answer was, "What this time?" Chickens in small numbers have endured in small pens behind garages in big cities, in old horse stalls on family farms, and in backyards in small towns all across this country for centuries. This is a real role for the domestic chicken and has been since before the time of recorded history.

Chickens became a business of any great size just over a hundred years ago and an industrialized commodity in less than half that time. The small chicken flock is an economic, cultural and dietary institution of long standing. In some cultures it has even had strong religious significance. The aberration is the chicken in rigid confinement.

Some of those sitting around me at that poultry market had sold literally thousands of chickens in their lifetimes. They had owned scores of breeds. Around the circle that day were breeders with Chanteclers, Saipans, Rhode Island Whites, Buff Cornish, La Fleche, Exchequer Leghorns, Nankin bantams; and pigeons, ducks, geese, and turkeys. Yet most would admit that it is the red and the barred hen that continues to anchor it all.

Those Rocks and Reds of our grandparents were the stuff of legend and most of us still slow each time we pass a little chicken yard hoping that some of that old blood has made a last stand there.

The real issue for the moment and perhaps for all time is not the birds, but how a greater number of the birds are now being produced. As I write this, it has just been announced that four out of every five birds tested from American grocery stores carried some type of bacteria. That ratio was found to be the same even in birds labeled "organic." Those are largely corporate produced birds that pass through mechanized processing that includes what some have come to call a dip in a "bacteria bath" of chilling water.

Still, these large producers point to the small independents, whose birds aren't in the stores, as a threat to their way of doing business and its vaunted concept of bio-security. Most of these large production chicken gulags are one power failure away from total carnage and they are delivering a product now being questioned by both classic chefs and purveyors of fast food chicken.

I fear that chicken of real quality and chicken in volume will always be concepts that are at odds with each other. When producers start piling up chickens you also start piling up manure, viscera and feathers and all of the problems associated with crowding and stress.

All over the country now, chickens into the hundreds of thousands are massed, just waiting for health, environmental or weather-related disasters to befall them. They are no longer chickens by nature's design, but eggs and meat by corporate dictate. The birds are just viewed as one of the means to that end.

Already under way are experiments that have real muscle tissue (meat) growing to serving sized segments in petri dishes. No flu could overcome something such as that. Nor would you or I want to eat it, either.

Cheap chicken from inside cold steel walls now fills wraps and cardboard buckets, tastes chicken-like, and does meet basic protein needs. Someday it may even be irradiated to the point where it is germ-free. It is a product, but a product devoid of any real pleasure in its consumption, personal rewards in its production or even fair pay for the workmen who labor in the wastes and stench to bring it out of those steel walls.

It isn't the small flock producers who have compromised chicken production. They are the ones endeavoring to hold fast to the many old breeds, the traditional practices, and the production style that will maintain true integrity and sustainability in poultry production.

I once commented to a Society for the Preservation of Poultry Antiquities member of great zeal that if we were successful with our preservation efforts we would render all such organizations obsolete.

The preservation breeder/producer of today should aspire to be a good purebred producer of the future. Whether with a flock of twenty birds or two hundred everyone should be an engine of quality. If it is novelty you seek, collect old mustache cups. If it is a speculator you aspire to be, then put your money into stocks and bonds and corn futures.

Whether your landbase is measured in square feet or acres, these feathered wonders were originally nearly all bred to be farmers' fowl. Returning them to that role will be the way to serve them best.

The hardest thing to do with these birds now may be to shake off the hobby fowl image that clings around them. That will have to be changed one meat bird and one dozen eggs at a time.

For most, this is still seat-of-the-pants stuff. These are businesses begun with the sale of extra eggs to friends at church or people asking if there are any extra to sell after learning you have been raising your own meat birds. Heirloom birds do now have a bit of panache showing up everywhere from Martha Stewart's opening credits to the Wall Street Journal. Range broiler themed articles have even shown up in the traditional agribiz, slick paper magazines. They aren't just our own little treasures any more.

On the other hand, individual producers are still their primary source. Maran and Welsummer chicks did not appear in any U.S. catalog of which I am aware until 2005. By that time they were already old hat at our local farmers' market. Their dark brown eggs are now featured items at a number of farmers' markets across the state. A couple of years ago eating even one of those precious eggs would have been a nearly unthinkable act.

Niche poultry markets emerge and are often filled within the blink of an eye. Let a high profile TV chef start singing the praises of large white eggs and Minorcas and Leghorns could break through anew. The real traditionalist will say that far and away the best tasting eggs are those laid by Standard American Pit Games. Historically most important may be the parchment colored eggs of the Java.

There must be an acceptance of just how finite some of these markets can be and just how much can be done with certain of the heirloom breeds.

When we had Sultans we might just sell twenty to thirty chicks a year — the output from a single, highly fertile pair. Several of those went to just two people wanting to start small flocks of their own. Nearly as many were sold two or three at a time to folks wanting some of those cute little "top knot chicks with all the feathers."

This was not the fault of the birds in this line, but their niche was truly tiny and the multi-colored Polish owned the crested bird trade in our area. It was our belief that our time and resources would be better rewarded if given over to larger, more productive birds with greater appeal to the smallholders that make up the largest share of our market.

What I am sure of now is that people are already at work developing better egg-laying strains of the Polish breed, better yielding White Rocks, and a New Hampshire to perhaps rival the Label Rouge broilers now so popular in France. Ponder for a moment just what could be done now with a flock of artfully bred Buff Plymouth Rocks, Black Wyandottes, Barred Leghorns or Light Sussex?

A successful small farm is a quite diverse smallholding. It is planned to produce income at every season and every week if possible. Into that scenario few ventures fit as well as one or two poultry ventures of modest scale. A small farm should never be too dependent upon a single species or venture. When a farm, whatever its size, becomes too narrow in its focus, it becomes too dependent upon forces beyond the abilities of the individual producer to control.

If farmers feel they have a real connection with poultry production then trial small lots of other poultry species that would complement existing ventures in place on the farm. Heritage turkeys have been one such venture for us. Two years ago one of our better Royal Palm pairs produced a total of fifty-five poults from spring into late summer.

Alternative poultry ventures are many and varied and most smallholders should find one or two with some appeal. A good many will even be in keeping with breed preservation work.

ADDITIONAL POULTRY VENTURES BEYOND CHICKENS

There are numerous poultry ventures, in addition to chickens that are worth looking into.

Consider small flocks of pigeons. They may be white Homers for release at weddings or other special occasions, colored Homers to be used for dog training, ornamentals like Fantails and Parlor Rollers that are real people pleasers at the swaps, and a few Kings or Mondaines for squabs.

Gamebirds fit a real niche where state licensing measures aren't too prohibitive. Check first with your local Conservation Agent. Here in Missouri a fifty-dollar game breeder's license enables you to propagate and market gamebirds such as pheasants and quail. Non-indigenous species do not need licensing and the accompanying on-farm inspections.

"Ornamentals" have found a niche with the growing number of "rurbans" — rural urbanites that want a few distinctive birds to function rather like animated flowers.

Ornamentals is a very broad heading and can include everything from peafowl to brightly colored pheasants to swans to fantail pigeons. These are often high dollar birds, may reproduce only in quite modest numbers — often not beginning until the age of two or three, and may require special care. Experience with peafowl, for example, will divide people into one of two camps: 1) how beautiful or 2) how loud and annoying.

Peafowl are now bred in several color varieties, but a pair will not produce with any sort of consistency until their third year of life. Still there are several farms in Missouri alone with hundreds of peafowl on hand. At these you will find pairs of some of the rarer colors selling for hundreds of dollars even when quite young. Swans are similarly big, beautiful, and mean tempered.

A pair of prime breeder, three-year-old swans can set you back one thousand dollars or more per bird. They will require swimmable water of at least five feet in depth for successful matings to occur. Cygnets, odd birds, and birds bought just ahead of cold weather can sometimes be had at substantial savings. I believe swans may find a growing market outlet, as they will drive away nuisance Canadian geese from ponds and small lakes where they will brook no intruders on their watery turf.

Ornamental pheasants like Reeves, Red Gold, Yellow Gold, Silver, and Lady Amherst's have a long history as aviary birds. Their colors are quite striking and of the group, the Golds are perhaps the easiest to tend and propagate.

In a number of rural lifestyle publications of late I have seen mention of and plans for stand-alone dovecotes and those that are design elements of outbuildings. In olden days such units produced squab for the farmer's table. Rollers and Homers are probably the best choices for those wanting flying birds and measures must be taken to prevent bird losses to hawks.

The guinea is perhaps the wildest of the domestic fowl. The Pearl variety looks very much like its wild ancestors still to be found on the plains of Africa.

Guinea have a well-earned reputation as feathered watchdogs and are believed to keep lawns rid of a great many different insect pests. Old timers say that they even drive away snakes.

They are bred in many colors and fried young guinea is a Midwestern specialty. They yield all dark meat and a "French" variety now is said to weigh in at one to two pounds heavier than other varieties. Still, it is their reputation as a tick deterrent in Lyme Disease areas that has focused most of the recent attention on them.

There are big ducks (Rouen), little ducks (Call), egg-laying ducks (Khaki Campbell), and some that fit no easy category (Crested). I think it is safe to say that there is a duck for just about everyone.

Once past the duckling stage they are one of the most hardy and self-sustaining of all domestic poultry varieties. The Muscovy produces breast meat many have compared to fine roast beef and other duck breeds that lay like Leghorns. They need simple housing, most don't need swimmable water, and they are generally quite prolific.

Geese have gained in interest of late with the rediscovery of small populations of Cotton Patch Geese. This variety was developed to clear weed growth from fields of growing cotton. Weeder geese (generally Gray goose crosses) are still used in some types of fruit production.

Two auto-sexing breeds, the smaller Shetland and the Pilgrim, have also gained renown as heirloom breeds that are relatively few in number. In fact, all but the Emden and Gray goose varieties have some sort of minor breed status.

Geese have value as meat birds and ornamentals and are grazers in season. It would not be a Dickensian Christmas without a roast goose on the table.

Many of the Standard poultry breeds have "mini-me" counterparts, bantam versions of themselves. There are also a number of varieties (Rosecombs, Sebrights, etc.) that are only bred as bantams.

Bantams have mature weights of less than two pounds and most are bred as a treat for the eyes and a challenge for the breeder's arts. Three bantam eggs can replace two Standard hen eggs in most recipes.

There is a chicken variety, the Silkie, which actually falls somewhere between bantams and standard-sized fowl. They are now bred in several colors, but for marketing purposes the Silkie variety of choice should be snow white with a deep mulberry colored skin.

Some Asians value these birds quite highly and even accord their black flesh certain medical qualities. I have a friend in Pennsylvania that sells several lots of four hundred to five hundred Silkies each year to the Chinese restaurant trade in New York City. He receives well over four dollars per pound on foot, but the birds are at least twenty weeks old before reaching any sort of handy weight. They are not a very productive variety and eggs must be removed quickly lest the little hens go broody. In parts of Europe, robins and lark-like birds are farm raised for sale. No, I'm not advocating that practice here and I'm not condemning it either, but do let this stimulate your thinking. What have you seen or read about in the way of fowl production that could be given a trial on your farm and offered to your markets?

Could you sell pickled eggs from a small colony of Coturnix quail? Does a shooting preserve near you need Chukars or flying Mallards?

At one time the keeping of what we now term the heritage breeds prompted no less a poet than Robert Frost to pen a poem about hens and their care. In their care he saw much virtue and even art.

That to me, is what the heirloom breeds embody, the art of the breeder and the passion of the agrarian heart. They haven't become static creations like the production hybrids so hammered and tightly fitted by today's industrialized agriculture. Within the parameters of the heirloom breeds they can still be bred to fit different farms and poultry yards. These are the breeds that can rise to any challenge thrown at them by shifting environments, climates and/or economies.

I have two friends now working diligently with the Rosecomb Rhode Island White breed. One is working to make her flock better layers. The other sees in their type and growth pattern the potential to develop them into a good meat bird. That kind of genetic flex can now only be found in the remaining populations of heirloom breeds.

Granted, we will never see any Fayoumis broilers or Cochins setting laying records, but all of the heirloom breeds can still be said to have untapped potential. There are still great amounts of genetic variability for the performance traits in many of these breeds.

I can remember Barred Plymouth Rocks from a local hatchery that had barring as crisp and even as if it had been painted onto the birds. Feet and beaks were a bright, clear yellow and they were big yet practical in their type. Where are they now?

Earlier I mentioned my own decades long search for the big outline "English" White Leghorns my grandparents once had. They were bought every few years as chicks from a local hatchery in far northern St. Louis County, Missouri. Did they simply disappear in a flurry of post-World War II suburban development? If so, why did my friend in eastern Kentucky grow up hearing of "English" White Leghorns, too?

Something no one would have thought even possible fifteen years ago has come to pass in just the last few years. A number of small, family-owned hatcheries have come into being. For the most part their entire stock and trade is in the heirloom breeds. A hatchery producing a single new breed, the Braggs Mountain Buff has even appeared, bred and developed for the family farm.

There is a new energy at work behind the scenes. There are enthusiasts working alone or in small groups and they are starting to make things happen.

Now there is a very real need for a commonality of voice for and from these people. They have the numbers, economic clout, quality matters, and consumer support that give them greater strength and position than perhaps even they realize. These birds, these people, and the traditional production practices that they employ are all what the consuming public really wants in the poultry products for their families.

These small, family-owned hatcheries have created positive images that the corporate types either try to co-opt for themselves via contract production or hide behind in veiled advertising campaigns. If Tyson or Perdue are family farms then pigs really can fly.

I have to believe that much less would be made of Avian Influenza and similar matters if the media had somewhere to turn to hear the voices of small, independent producers. No, we don't need our own PR flacks. We do need family farmers telling their own stories in their own words and seeking meaningful dialogues with the consuming public.

An *E. coli* fueled panic of even a few weeks length would cripple the high volume, commercial sector. Quickly, they would respond by turning off incubators, dumping eggs, and even euthanizing newly hatched chicks and young birds. However, bird and cash flows would be disrupted for months afterward. If word of the chick destruction got to the general public it would be a real image killer.

The positive image small producers have doesn't make them immune from such problems, but consumers don't look to them every time expecting the very worst. Consumers do, however, need to hear more from this particular farming sector and be better educated about how they are being fed now.

What will give heirloom producers the greatest recognition in the farm sector would be to fully and clearly establish their economic "bona fides." These people have been off the radar for more than fifty years and perhaps this has created a greater memory gap in rural America than with

POULTRY SUCCESS OVER LONG DISTANCES

Overcoming the challenges of working over long distances, Braggs Mountain Poultry has developed a system to successfully have their poultry breeding in one state and the rest of the operation in another. Over a distance of 350 miles, requiring a non-stop 6-hour drive, David Andrews and his partners at Braggs Mountain Poultry have divided the operation so that the hatchery operates near Whitney, Texas, while the headquarters is based in Fort Gibson, Oklahoma. The headquarters processes orders for baby chicks, launches advertising and develops roosters. Meeting halfway between the two locations, by coordinating schedules, to transfer their breeding roosters have allowed Braggs Mountain Poultry overcome the distance challenge.

Braggs Mountain Poultry has been working diligently for over 26 years to breed and refine a hen that lays an extra large brown egg. After working on what they named the Braggs Mountain Buff they now have a "golden hen that lays the jumbo brown egg." The early breeding involved 9-10 different breeds to craft the desired traits. With a vision for the future in mind and after 15 years of breeding work, the Braggs Mountain Buff was finally stabilized and now reproduces true to type.

Here is the recipe that has resulted in the Braggs Mountain Buff chicken:

Ingredients:

40% Rhode Island Red (or derivatives of Rhode Island Red such as Production Red, Cherry Egger, Red Star or New Hampshire Red)

20% Buff Rock

10% Black Langshan

10% Light Brahma

10% Buff Orpington

10% traces of Barred Plymouth Rock, Jersey White Giant, White Leghorn, and Black Australorp

Instructions:

Spend the next 18 years picking the handsomest and most vigorous individuals for breeding purposes and breed them for the golden buff feather color.

Next pick only the heaviest, largest and highest quality eggs for the incubator. Hatch these for your breeding seedstock.

Repeat.

The present day Braggs Mountain Buffs are a good dual-purpose bird. Good both for laying eggs and meat production. The bird is a large bird with light yellow skin that produces big brown eggs. In addition, the birds are very gentle — and this includes the roosters. Many repeat customers have commented to David Andrews and his partners about how docile the birds are on their farms.

At Braggs Mountain Poultry, the birds are raised free range on pasture and are shut up for protection during the night. Braggs Mountain Buff birds have shown to have good heat and cold tolerance.

Braggs Mountain Buffs are selected for longevity and David Andrews doesn't do anything special to pamper his birds. He feels that a better bird is produced and the flock grows stronger without any pampering. Toughening the birds at Braggs Mountain Poultry means birds are more able to withstand the potential pressures of their new environment.

For example, David advises customers that new bird arrivals be taken out of the shipping box and their beaks dipped in sugar water. This provides energy and encourages them to drink water. In a "do as I say, not as I do" situation, David doesn't dip his chick beaks – just lets them fend for themselves — a kind of survival of the fittest plan.

David takes pride in his durable birds. He likes customers to have a confidence that the birds have been raised tough. These birds are his "seedstock" and they need to be tough for the benefit of future generations. He raises the pullets in a tough environment so that their offspring will be better able to better tolerate the conditions at their future home.

Hatched Braggs Mountain Buff chicks are available from early February to early June as straight run (unsexed) birds. They are sent by Priority Mail through the post office to anywhere in the continental United States. The straight run birds usually average 50/50 pullets and cockerels. Presently they only sell baby chicks, but

Braggs Mountain Poultry may branch out and ship eggs in the future.

At the hatchery, David Andrews develops their pullets and around Christmas, he and his partner meet halfway between Ft. Gibson, Oklahoma and Whitney, Texas to exchange roosters for breeding.

Braggs Mountain Poultry no longer breed Braggs Mountain White. They offered this alternative for one year but there was little customer demand, so the Braggs Mountain White was dropped. Customers are enamored with the Buff color. "In some cases the white is a better bird than the Buff," David says. Five percent of the Braggs Mountain Buff types reproduce as white due to a recessive gene.

David Andrews is the Braggs Mountain Poultry manager in Whitney, Texas.

Braggs Mountain Poultry can be reached at 1558 Kreider Road, Fort Gibson, Oklahoma 74434.

urban-based consumers. Iowa corn may have fed a lot of chickens, but they were all way far away in California, along the Del-Mar Peninsula or down along the Missouri and Arkansas border.

A flock of chickens may have half the number of legs of a herd of purebred cattle, but in all other ways they are no less their equal. Fifty black hens of the right breeding, in the right place could generate tens of thousands of dollars in yearly sales with the right buyers. My Pennsylvania Dutch friend, whom I mentioned earlier, gets twelve dollars or more per bird for his young Silkies when sold in four hundred bird lots. I'll let you make your own jokes about "chicken feed," but make them at your own peril. We went through a lot of years when the live chickens most saw were the few old hens that might pop up at the local sale barn. At the sale barn, they might bring only twenty-five to fifty cents each. Then, several years ago, at one of the first exotic animal auctions in our area, a crate of ten older Ameraucana hens was carried into the auction ring. The crowd was mostly locals in for the show, but bids came fast on the old girls and the gavel fell at eleven dollars per bird. The crowd grew steadily quieter as one-by-one they did the math. They had just seen a coop of ten chickens bring one hundred ten dollars, good birds had actually been selling for those prices for quite a long time, but not where the general public could see them.

Word of the $2.50 a pound range broiler at once thrilled and frightened a great many farm folk. Many rushed to see where they needed to go to deliver them up by the thousands. When told that there were no live bird buying stations in which to sell them and that the seller would have to do the processing and marketing themselves their ardor quickly cooled. This chicken was just too much of the wrong kind of work.

PROMOTING RARE & HERITAGE POULTRY BREEDS

In looking at rare and heritage poultry breeds there are many ideas that I would like to share that would be useful to get poultry breeds more prominence.

• One is forming a producers' league or alliance to give the calling a verifiable national image and voice. Needed is a national center with a lending library, information sharing network, demonstration flocks, market reporting, a news service, and a national publication. In the twenty-first century a 24/7/365 presence is needed to tell the small-scale producer's story.

• A regular mailing of press releases to national and local media giving the small producer's perspective is sorely needed.

• Any national group would need regular polling of its members with everyone given an equal vote to determine future direction.

• A presence on the Internet including a fact checking service would help get poultry breeds more prominence.

• There is a need for a computerized system to facilitate the exchange of genetics between cooperating producers. It could even help to coordinate the transportation of the needed birds.

• A program to assist with pedigreed matings and document bird purity is warranted if it doesn't become overly intrusive or costly.

• The emerging egg and meat bird sector based on the heirloom breeds needs to be recognized for the growing role it has to play.

• There is also a very real need for a number of national and regional events where the heirloom breeds can be promoted and marketed. Not bird shows, but bird conferences, with teaching seminars and even trade shows.

I, like most small farmers now, am more than just a little leery of efforts to over-organize things. Let three folks meet over coffee now and someone from the government will swing by to organize them into some sort of coffee drinkers' cooperative.

Many producers now are forming very simple alliances or working associations to create a better market presence and to utilize the different talents of the members of the alliance. One might have computer skills, another writing talents, and another driving skills and together they can do a better job of buying, producing, and marketing.

Joint effort at many levels has a role to play in poultry breed preservation work. It can range from two friends working together to develop a viable breeding population of a single breed to a dozen or so folks coming together to market table eggs or launch a small hatchery. Cooperative effort that becomes too constrictive and rule heavy is worse, far worse, than independent producers continuing to work alone.

What happens after making the start, getting through the numbers building stage, and at last, having a group of birds in goodly numbers?

What then? What's next? Should you sell them all and take up a new challenge like salsa dancing? Well, really, yes.

No, don't sell them all and go chasing something new, but do realize that you and they are now ready to set off in an entirely new direction, facing new challenges. The best way to begin may just be to cull away the bottom third of the birds at hand.

With this one stroke you will amp up your flock on nearly all levels. If it can be done without harming your ability to serve your existing markets nothing else can do so much to make you a better producer.

It isn't so much a matter of simply paring costs, but rather freeing time and resources to do a better job with the better birds that remain. The quest to make a flock better should be a never-ending one. The more quality improves, the more difficult it can become to continue to ramp up flock performance.

It will be far easier to turn a flock of 160 egg per year layers into a flock of 175 egg per year layers than to turn a flock of 200 egg birds into a flock of 210 egg birds. The same will be true as producers go about shaving days of age off of days to market. Still, there should never be a "good enough!"

It is conceivable that at some point, maybe even five to ten or more years in, circumstances or dwindling genetic strengths may compel producers to start over completely. Some may even welcome such a challenge. In thirty-five years with purebred hogs we took apart and rebuilt our Duroc herd three different times and also worked for a time with two other breeds.

The challenge of hewing to an ever-higher standard will be greater with some breeds than others. The Barred Plymouth Rock with their distinctive color pattern and yellow beak and legs will give many challenges when striving to build them into an ever more prolific farm flock. It will not be as great of a challenge compared to working with the Orloff or Redcap breeds. There the first issue to be resolved is just how much can be legitimately expected from them as breeders progress in certain directions. The goals for any breed must be tempered by the realities of what they are now (what genetic pieces remain) and what they were bred to do. For some, job one was simply to be moderately productive in a very harsh environment. A ten egg per hen gain with them may be a victory of epic proportions.

Some may even just need to be held in place and kept just as they are. To make them uber productive, you may have to take them apart too much. Even a silver saddle on a sow won't make her step out like a Tennessee Walking Horse. Long Crowers, Frizzles, Yokohamas, Phoenix, and a good many more were developed and long time bred almost entirely for aesthetic reasons. There is nothing wrong with this and something commendable for those who take up a living thing to make it ever more attractive and compelling.

Your choice of breed or breeds must very much be your own. Their appearance is every bit as valid a selection factor as any other. The two breeds that first caught my eye as a youngster were the Rhode Island Red and the Barred Plymouth Rock. I have owned small flocks of them both on and off for decades. It took several years to find the strain of Rosecomb Rhode Island Reds with which we are now working. Red is, fortunately, my wife's favorite color.

I have a friend, a former cockfighter raiser, who cannot pass by any bird with a history as a scrapper. He has a real tender spot for the tall, primitive looking Asian breeds like the Aseel and Saipan. Their fearsome appearance is off putting to many and the practical purposes to which they can be put will certainly take an imaginative sort of producer. Perhaps one day, it may be within them to lay an egg with gourmet properties or produce a most distinctive roasting bird.

The selling process for some of these birds is going to be every bit as challenging as getting them restored and propagating abundantly. Yet with them you will not have to come across as some sort of used car salesman in overalls. Basically, all that you have to do is share with others what sold you on their breed or breeds of choice.

Some years ago at our local farmers' market, I sold pork sausage made from a surplus male Mulefoot hog we had raised that year. My little sign informed potential buyers that their purchases would help to preserve a rare old breed of swine, one of the very oldest in the United States. The very first lady who stopped by pointedly wanted to know how eating one of them could help preserve them. That was my foot in the door and, bless her heart, she got my whole sermon.

PROMOTING A RARE BREED

When asked by customers how consuming a rare breed can help preserve it, I offer the following tips of explanation.

• When someone buys your eggs or poultry meat they are providing you, the committed producer, with working capital and restoring the traditional demand for these creatures. Without a market to economically undergird them they would decline farther. With their purchases they are investing in them, too.

• This is the role for which these breeds were developed.

• Every year surplus males and females that are not of breeder quality are produced. Likewise, eggs are often produced far beyond the traditional hatching season and in greater numbers than any demand possible for live birds. This is the most conscionable and appropriate use for them.

• Only a certain segment of each year's crop is ever of the quality to be considered as retaining for seedstock.

• Put this into a one page handout, fit it on the back of a business card, use it as the theme for a newsletter issue, put it into a story for your local newspaper. Hey, wait a minute! Did you just create your own advertising campaign? Not as tough as you thought, was it?

• Nothing stays the same in this farming life of ours. Change is inevitable even down on the family farm.

Let me again use our own small efforts as an example. I believe our goal of producing and promoting heirloom breeds with strong family farm ties is a sound one with the legs to go the distance. The breeds with which we work have not always been the same, however. In fact, we have dispersed a number of breeding flocks in just the past few months.

Part of it is that I am getting older and there does seem to be fewer hours of sunlight into each and every day. Every one of us must determine at what level of operation we are most comfortable and content. I find myself standing often in the poultry yard wondering if I should focus more resources with this breed or am I really doing all that I should with that group over there?

We are now well down from a one time high of twenty-three breeds and varieties. Some of those really hurt when they went to new owners and others, I had to admit, would be better off in hands other than mine. With some of the very rare breeds you can become very hesitant to step away, but no one can do it all.

I've had too many birds handed to me with the words, "you will know what to do with these." Sometimes the best thing to do with them is to turn and hand them to another.

I expect that, as time passes, our breed count will continue to go down. We are steadily trending toward fewer breed groups, but more birds in those groups. Five breeds and a couple of experimental groups looks like a good numerical goal for us now. Many of the most successful breeders I have encountered have worked a lifetime with just one or two breeds. The challenges never lessened for them and their accomplishments continued ever upward.

I suspect certain new birds may always tempt me. There are at least a handful of breeds that could still excite and draw me out. Among them would be Yellow Hungarians just because of their near mythic status.

I'll always be open to good Wyandottes, and, of course, someone somewhere still may have some of those "English" White Leghorns.

The small farm flock, typically of four hundred birds or less, is a rural icon that could be coming around again. It may now number fifty or even just twenty-five birds, but the working chicken flock is hitting the time clock again.

Will it become as important as it was seventy-five to eighty years ago? To some of us it will.

The small farmer and the small poultry flock are tied together about as closely as anything to ever emerge from production agriculture in the

United States. It was even a theme in the movies of the thirties and forties, television series in the fifties and sixties, books in the seventies and eighties that fueled environmental and back-to-the-land movements, and now there is a poultry renaissance that has created new opportunities for a new group of small scale, human-scale farmers.

There is a vigor, a fire for life if you will, in the heritage birds. It is how they came through so much neglect and oversight to reach this point of new promise. Some may have lost ground genetically and productivity-wise, but through no real fault of their own.

A trio of such birds or a small box of their baby chicks have in them still all manner of promise and potential. You will have to work to bring it out of them and to bring out of them all that it takes to be a successful breeder of them.

Whether your breed of choice is the old standby White Plymouth Rock, the sprightly Silver Penciled Hamburg or the venerable Buff Wyandotte, the future with them can and should be every bit as bright and rich as their past — if not even more so.

What they will be is up to you, the smallholders and breeders of today. Old cockfighters are familiar with the call to, "Pit them up." To those of you with an interest in the rare and heirloom chicken breeds the call comes echoing down to you from over the ages, "Put them back to the work for which they were bred."

HERITAGE BREEDS VS. NATIVIZED GENETICS

"Do you use heritage breeds?" The question, becoming more and more common, indicates the breadth and depth of the heritage breeds movement. As direct farm marketers hone their sales stories, this new product differentiation mystique needs to be analyzed realistically.

First, the caveats — I'm not a scientist or a geneticist. Goodness, I don't even know whether my aunt's grandniece is my second cousin or first cousin once removed. That said, I have great, enduring respect for the American Livestock Breeds Conservancy and bear no ill will to any gene pool conservationist.

That said, I find myself in a similar position to where I was nearly a decade ago when I finally began using the phrase "beyond organic" to answer the question: "Are you organic?" Too often I find a certain elitist condescension in the question, an almost implied: "What's wrong with you? Not good enough for us?"

To see the erosion of organic standards, the acquisition of organic labels by food multi-nationals, and the general industrialization of the term, if not bastardization, I'm more satisfied than ever at finding a counter-term that was both positive and explanatory. Something that stimulated conversation instead of generating hardening of the categories.

Here at Polyface Farm, in general we do not use registered heritage breeds but rather than mumbling that embarrassingly into my hand, I will offer a progressive alternative: we promote nativized genetics. For example, we use non-hybrid Black Australorps, Barred Rocks, and Rhode Island Reds for laying hens. We don't use the hybrid sex-links like Dekalb Golds, Golden Comets, Cherry Eggers or any of the other Leghorn/Rhode Island Red crosses. The old-fashioned heavy breed layer has been the linchpin of my tangible appreciation toward heritage breeds.

Using the three breeds allows us to amalgamate older and younger birds into large flocks and still be able to track which ones are older or younger. For some reason, chickens don't take ear tags very well. Each time we purchase pullets, we get only one of those three breeds and since we get new chicks two times per year, this rotation spans 18 months.

In the last two years, however, we've been unable to obtain the numbers for pure initial flocks. The last time we purchased Australorps chicks, for example, it took 8 weeks to get the full 2,500 pullets. The hatchery said we overran the entire U.S. breeder flock capability — *i.e.* not enough breeder hens exist to supply enough eggs at one time to produce 2,500 pullets in one hatching.

Lest anyone immediately label me an empire builder for getting this large number at once, compared to the industry, this is a tiny number of birds. Yes, it's a commercial scale, but it's a spit in the ocean compared to populating an industrial confinement house. And if we really want to displace the industrial food system, we need to scale up beyond our backyards, as important as those may be. The future needs all the backyards plus some commercial-scaled producers.

Chicks coming in over this long a span is a management nightmare for us and so now we're faced with getting only Rhode Island Reds, which still have enough non-hybrid breeders to produce the number of hatching eggs at one time to ship a batch of chicks this large. Several years ago as we began overrunning hatchery capacity we hoped that other pastured egg producers would step up and order these birds in numbers large enough to encourage breeder flock operators to expand their genetically-endangered flocks.

But nearly all the pastured layer flocks I know of — and certainly the commercial-sized ones — are using sex-links. The hoped-for heavy breeder flock expansion did not and is not occurring. One reason we use the poorer laying heavy breeds is because the eggs are better when the production doesn't extract such a heavy toll on the bird's energy. By going to a bigger body and fewer eggs per year, the hen stays calmer (no blood spots), healthier, and transfers that excess nutrition into the egg. She's also smarter. For layers, what we need is to isolate the hens that lay the darkest-yolk eggs since yolk color indicates aggressive scavenging — earthworms, grubs, grass, etc. If we would consistently breed those in a couple of generations we'd get rid of the loafers that just sit inside the eggmobile lounging around the feeders. In any group of animals, certain individuals are workers and others are loafers. We need to select for those that get out there and forage.

Now we move to broilers. For two years we tried non-hybrids but found them too difficult to sell. Razor-breasted, dark-meated, tougher birds priced embarrassingly high did not find enough market acceptance, even among our Weston A. Price folks, to make the endeavor worthwhile. The taste, texture, and nutrition of the older birds, I'm sure, is superior. But at the end of the day, we must be able to sell enough to pay the taxes and put shoes on our feet. For meat birds, why can't we look at fat color as the foraging indicator? In any batch of 1,000 pastured broilers we process, a certain number have decidedly darker colored fat without any apparent change in physiology. In other words, they still weigh as much, have a heavy breast, and in all appearance look just like their industrial genetics pattern, but clearly they've been eating much more grass than their counterparts.

If we would select those parents over several generations, we would soon have our cake and eat it too. We'd have nice consumer-acceptable double breasts, but we'd also have aggressive foraging. In a wider sense, we could even select for birds that run when hawks come. As a pastured poultry producer, my ideal chicken is easy to describe, so why can't we select for that and develop those traits as a diversified genetic blueprint? Is that not as noble as keeping a flock that somebody else in another part of the world selected a century ago to exhibit the characteristics they deemed important in their day in their area?

I would love to see genetics backed off about 10-20 percent from the industrial birds so that we could have something with a heavy breast that doesn't grow quite so fast, grazes more, and is more active. The very slow growing heritage breeds, at least in my experience, simply do not have enough marketplace acceptance to create a viable business scale. It's kind of an all or nothing situation. You either go with an extremely different phenotype or a completely industrialized one. I'd love to see some middle ground.

Because every generation in a certain locale carries genetic memory, the adaptation to place occurs in plants as well as birds. Nuances of climatization occur throughout the biological world. In fact, this is the foundation for heritage breeds. That's why most breeds carry a geographic component in their name.

The point is that the breeds we've collectively chosen to label as heritage originated someplace else, someplace with unique climate, latitude, geography, and diet. Whatever genetics functioned best in that locale expressed itself in the phenotype, disposition, and production character of that critter. In nature, function always dictates form.

Wouldn't it be better for each of us to begin systematically and carefully building nativized genetics in our flocks? Why stop the adaptation clock in time and space? Why not continue the process here in our own counties, our own food sheds, and our own bioregions? Over time, this would actually add diversity to the gene pool rather than limiting it to what was a century ago, or what is today.

If the gene pool we currently enjoy is the culmination of this nativization process, why not let the process continue? Why not encourage it? In the future, we could have a hundred new American breeds bio-regionally specific. Would that not be as valuable an addition to the gene pool as the current old country base?

In this context, let me offer a philosophical argument to the folks who say, "My breed is the best." If that were the case, are we supposed to let all the other breeds go extinct? I mean, don't we all want to have the best? It's philosophically reprehensible to me to suggest that none of the other breeds except mine have a place at the table. What happened to inclusiveness? What happened to diversity?

I think when we market our specific breed as the be all and end all — certainly chefs are guilty of this too — we practice genetic elitism that if carried to enough converts could jeopardize genetic diversity at its base. In truth, some are better than others. Finding functionality and balance for our own operations will require observing many characteristics, weighing the subtle differences we see as we measure litter size to heat tolerance, or carcass yield to grazing ability. An old folks saying expresses it this way: "More differences exist within breeds than from breed to breed."

I've come to believe that if we tout our heritage breed as our differentiation, if that becomes our stock-in-trade, it actually feeds the Bambization and Easter Bunnyzation of domestic livestock. If you want to use a heritage breed, that's great. But think carefully about the pet creation ramifications when your selling point is that they are cute or fluffy or show quality. That doesn't help anybody eat. And ultimately, the biggest danger to poultry genetics is failure to serve them at the table, because that is what really creates the need to maintain and reproduce.

This is why now when asked: "Do you use heritage breeds?" I simply respond good-naturedly: "Oh, we're creating some by using nativized genetics." Then, instead of ending a conversation, it starts one. And isn't that a lot more fun in the long run?

Now let's go out and get some more breeds into existence.

— *Joel Salatin, Polyface Farm*

Originally published in Acres U.S.A.

• CHAPTER 10 •

POULTRY BREEDS — THE FINAL WORDS

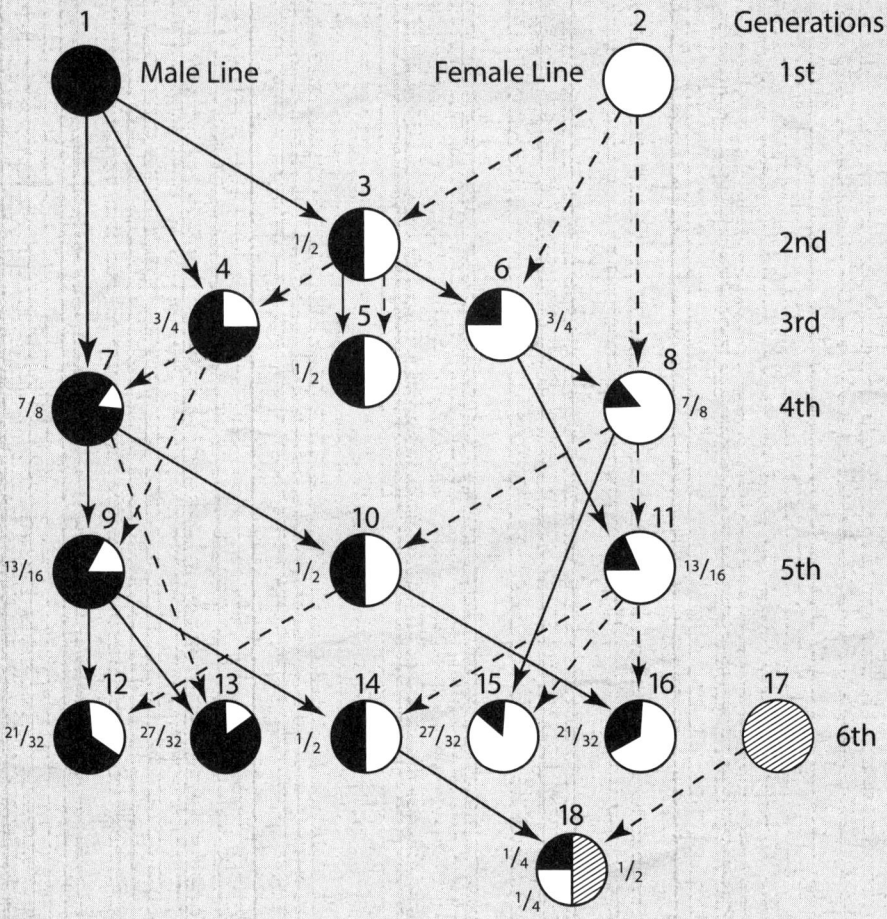

LINE BREEDING CHART — *The circles represent individuals or groups of individuals used in the breeding or resulting from the matings made. Black represents the male line and white the female line, the relative amount of each color in a circle as well as the fractional figures at the side indicating the relative amount of each blood carried. The large number at the top of each circle is given to it so as to make the discussion of the matings in the text clear and easy to follow. A solid black line connecting one circle with another indicates that a male from the first group was used in the mating to produce the second group, while a dotted line indicates that a female was used. The shaded circle at the lower right hand corner indicates the introduction of a third blood line, and circle 18 shows the proportion of the three bloods in the resulting offspring. (After Felch, Pierce and Lippincott.) From The Mating and Breeding of Poultry, Harry Lamon and Rob R. Slocum.*

ONE LATE WINTER AFTERNOON A FEW YEARS BACK I ENTERED THE HOUSE AFTER DOING LATE CHORES WHEN THE VERY LAST DAYLIGHT WAS LEAVING THE SKY. Chores had taken over three hours and Phyllis asked why I was so late to arrive for supper.

In cold weather I like to take my time with late chores and make sure that every bird gets something to eat, has had a good drink, and is well settled in before the long winter night arrives. That evening I dropped into my old chair with pad and pen in hand and a few minutes later looked up to tell Phyllis that we then had nearly two dozen chicken breeds and varieties on hand.

Some were just in pairs and trios, but as word got out about our work with chicken breed preservation, birds just seemed to find their way to our house. More than once at a farmers' market I would be standing to one side talking with friends, and a cage would be set at my feet with words to the affect of, "Here, you need these."

Having over twenty chicken varieties was just too many breeds to do them all justice and still get into the house in time for a hot supper. The next day after this revelation, we then began "the great cutback." We got down to fourteen breeds fairly quickly with some bartering and passing along to young people and those just entering the field.

A few months later we were back to seventeen breeds and the subject of breed choices became a real serious topic of discussion. When Missouri poultry folk meet at a bird swap or farm market, the question most often asked about is what did people notice bird-wise. Generally this question receives the standard answer, "Nothing I couldn't live without — still looking."

This is not meant as a criticism of an event or market, but is rather a roundabout way of saying that there are very few poultry raisers that don't still have a wish list of particular birds and will always have room for a special breed or variety. I am sure nearly everyone would find room for a few of the near mythic Hungarian Yellows (confined in North America to a few flocks in Canada), some seek new varieties of a favorite breed, others have a favorite bird class, and others, well, we all have different likes and dislikes.

What makes a breed special or interesting can be hard to nail down for some, but to the above list you can add those who keep and breed birds for a specific trait such as extra dark brown eggs, those working with rare colors or creating a new color for a breed (for instance the growing interest in lavender and lavender lacing), and rare and endangered breed raisers. On our farm, the focus has been on breeds that have a long association with the family farms of the Midwest. Over the years we have owned breeding groups of many of the breeds in this book and some of the first Welsummers and Marans in modern times were here in eastern Missouri. We are fortunate in that a good number of rare and heritage bird producers are found in the counties of east-central Missouri. The local breeders' group operates a series of farmers' markets and poultry themed special events at which birds as rare as the Rhode Island White and Welsummer are regularly seen and offered for sale. As of this writing we are working with a group of fellow modest sized producers to launch a cooperative hatchery that will be producing heritage breeds both in a variety of colors and patterns and with an emphasis on their more utilitarian aspects. We are very much about returning these great old breeds to their practical roots and building flocks to be truly viable pools of breeding stock for them.

Here, after a long, long winnowing process, including working with several others to create aligned flocks, we are currently working with just three breeds. They are the Buff Orpington, Rosecomb Rhode Island Red, and the Single Comb Rhode Island White. While I still have my own little wish list, perhaps I should share a bit about why we chose these three particular breeds.

The buff color in chickens has long been quite popular. It could be said that it was the presentation of a small breeding group of Buff Cochins to Queen Victoria that sparked a lasting interest in improved poultry breeding, exhibition, and resulted in greater organization and vision in the poultry sector. Over the years we have kept Buff Orpingtons and Buff Rocks and Buff Wyandottes. Early on we had trouble finding a Buff Orpington line that would perform the way we desired on our small farm. Some didn't reach good size, others lay poorly, and some produced quite small eggs. The Buff Rock and Buff Wyandotte both presented as possible alternatives, but with the Wyandottes there were egg issues, too, and size issues with the Rocks.

What clinched the Buff Orpington for us was that most basic of reasons — economics. For most people the buff chicken is the Buff Orpington. At markets and farm shows, when we set up the Rocks or Wyandottes, many would gather to see them, but it was the Orpington breed that attracted the most attention. This breed was the one with the most positive associations for people, dating back, perhaps, to a flock owned by their parents or grandparents. You don't have to hit this old country boy over the head more than a half dozen times or so before he begins to see the light, so we got busy finding and breeding better performing Buff Orpingtons.

The Buff Orpington has both history and name recognition going for it. It does have white skin and shanks and while these are popular on meat birds in Great Britain and the rest of Europe, it is yellow skin and shanks that are the most favored here. The good buff color, the size of the better bred birds in the breed, and their quiet nature make them most attractive to everyone — from those wanting a handful of birds for a backyard flock to those building a modest sized flock for brown egg and roaster sales.

Buff Orpingtons were one of the many breeds favored for caponization when that was a more common practice. Many lines still produce good broody hens, and I have to say again — that color draws a lot of attention. The Orpington is also bred in blue, black and white varieties, but it is the Buff Orpington that is one of the few chicken breeds with the name recognition factor on par with the Black Angus cow or Clydesdale horse.

However, this demand potential hinges very much upon a bird that is both eye catching and practically oriented. After many attempts, we found a breeding line that does a pretty good job in both of these areas. They lay fairly well, grow to good size, retain the broodiness trait still valued by a number of small-scale poultry raisers, and produce an egg of good size with an even, light brown color.

We will have to continue selective breeding to keep these traits in place and in balance. We have been quite closely linebreeding this strain for some time with no adverse effects and plan to return to the source supplier for the occasional flock addition. We have also begun some side matings with better males from the line trial. These have been mated to one or two females from other lines. This is a first step in flock building and also gives us an idea of how breeding birds from our flock might work for others.

The Rosecomb Rhode Island Reds are also a breed with name recognition and have good demand, although the Rosecombs are much less seen than the Single Combed variety. They are also favorites of my wife, Phyllis, and my late grandmother. Many people have become so accustomed to "Production" or "Performance Reds" that to see purebred birds with the deep, mahogany red of well bred Rhode Island Reds almost sparks arguments.

We chose the Rosecomb variety due to its rarer status and for our simpler housing, which is still comparable to farms of a century ago, the rosecomb varieties have greater winter tolerance. The Rosecombs are not nearly as vulnerable to freeze damage and frostbite, as are the single comb varieties.

The Rosecomb Rhode Island Reds, being both distinctive in appearance and more rare, have created more buyer interest for our birds. One family about fifty miles away seeks to buy 25 to 30 as hatched chicks from us each year for their youngsters to develop as show birds for project work and to supply eggs for the family table. The breed has been known to have some fertility problems as the birds age; so we try to always keep good young males waiting in the wings, use males for no more than two seasons, and work with other breeders with related lines to sometimes exchange breeding stock.

One fellow breeder's flocks were begun with birds directly from our flock. Our foundation male is still vigorous and working in that flock. We went back to that particular flock for the breeding male last season and will again next year as he has performed quite well with a set of hens that were essentially his aunts and cousins. He is helping greatly in the area of comb size and structure

The Single Comb Rhode Island White is a breed not currently recognized by the American Poultry Association. They recognize and sanction the Rosecomb variety only. The Rhode Island White was developed early in the last century when the White Wyandotte and White Plymouth Rock were dominant breeds with large numbers of breeders and well placed supporters.

The white breeds were popular with farmers producing for the commercial trade and backyard bird raisers. At the time, a number of white breeds including the Lamona were being developed. However, many challenged the need for so many white-feathered breeds and varieties. Some say that the Single Comb Rhode Island White simply ran afoul of poultry politics and infighting.

It is held that the two Rhode Island White varieties (rosecomb and single comb) are not akin and they weren't bred out of the Rhode Island Red. They are both all-white birds with no black points. The Single

Combed variety has been used in producing some hybridized laying lines and they are, hands down, the most productive hens on our small farm.

They are a bit smaller than White Rocks, but do have the body shape of the Rhode Island Red. The line we are working with arrived here in a box of assorted chick varieties from a fellow small farm breeder and with a note that read, "I believe you will like these."

And like them we did. They lay earlier and later in the year than anything else here; lay very few small, pullet eggs; and quickly go on to lay quite large, light brown eggs. We've always had good fertility in this line and they have good heat tolerance, too, always a plus with our hot and humid Missouri summers.

Some years ago I passed this line on to a father and son breeder team due to the challenge of trying to hatch and brood five different white breeds. Not too long ago their breeding interests changed and they decided to let the birds go. We acquired the small breeding flock they had kept and a number of baby chicks with which to re-launch a flock here. I have some size concerns with this line, but they can be managed with selective breeding for improved size as the vigor and breed character have always been quite good. They are still quite rare and we are always on the watch for other producers and sources of seedstock.

This three breed combo now works well for us, but it should not be considered a universal ideal, some might rather have just one breed, others have more favored breeds and birds with which they have a good history, and there still are other breeds on our "possibles" list. Good Wyandottes will always draw my attention. Light Brown Leghorns have long been another favorite, and every couple of years I'll order in a box of assorted chicks from one or another of the heirloom and rare breeders to check out the "competition" and get some hands-on experience with new-to-us breeds and varieties. We ask for each breed in enough numbers to assure the grow out of a good pair or trio.

As poultry keeping has again grown in popularity and the economic importance of numerous breeds have grown and ebbed in popularity particularly with the Dominique, Java, Sumatra, Welsummer, Maran, Buckeye and Delaware. A cover picture on a poultry magazine can now even fan at least some short-term interest in a particular breed or variety.

All sorts of producers, including some high volume commercial hatcheries, have taken up breeds of the moment. The Internet is now rife with "pop up" hatcheries that arrange to fill orders with chicks drop shipped from one of the few larger hatcheries — some of which do hatch the year around. Some of these even share the same sets of breed pictures and artwork.

While good selling prices are a short-term benefit, too much popularity and exposure can also lead to a flood of birds of lesser quality and poorer breed type and performance character. A rush into production for merely quick profits can see a lot of birds of poor type, constitution, and even coloring dumped into the marketplace and the gene pool. This also seems to draw a lot of buyers with limited experience and lacking in needed selection and breeding skills. With some of the very rarest of breeds you have to make the start with what is at hand, but birds that only just somewhat look the part are not the way to do it.

The American Livestock Breeds Conservancy has begun redoing its efforts in behalf of some classic breeds since the fervor that grew up around some of the breeds may have actually done them more lasting harm than good.

Tips to secure
quality heritage birds

The following are some tips to help with the search for a quality supplier of rare heritage breeds.

• Locate the purest possible sources of a breed's genetics. The longer a particular flock's history is — the better. A few may even have vestiges of a performance background to share with buyers.

• Become a student of the breed, its history, and the practical uses for which it was initially developed.

• Acquire the best possible genetic pieces with an eye for their vigor, growth in keeping with breed standards, and fertility.

• Begin breeding toward their traditional type and role. Cull ruthlessly while doing so.

I have recently seen Exchequer Leghorns, exotic colored Wyandottes, Delawares, and Javas offered by even some of the smallest of the commercial hatcheries. Some of these might be doing a quite serviceable job with one or another of these breeds, but I have heard some horror stories right now about what was supposed to be in a chick box in no way matching what actually arrived. And some of those boxes come from individuals and not just hatcheries. My take on things right now is buyer be careful, be very, very careful.

Breed selection must be a very careful, well thought out and executed action. I know the temptation is always present when opening a catalog or brochure to say I want three of these, two of those, and a couple of the black ones on the next page, but that works only if you're stocking a chicken zoo and not building poultry flocks.

Which choice of breed is certainly the question most asked at farm shows, bird meets and markets, and wherever those interested in poultry keeping are apt to meet. Breed choice could lead to the work of a lifetime — so it remains a very personal, very important long-range commitment.

I wouldn't tell anyone to take up a specific breed, but here on these pages, would like to offer lists of breeds that I believe would perform well under various circumstances. They are breeds long associated with specific farming practices and production facilities and that will meet the demands of various marketing segments.

LARGE BROWN EGG PRODUCTION BREED RECOMMENDATIONS

For large brown egg production, my list of breeds and varieties would have to include; White Plymouth Rocks, Black Australorps, Barred Plymouth Rocks, Single Comb Rhode Island Whites, Rhode Island Reds, New Hampshires, Welsummers, and Buff Orpingtons. Not every line within each of these breeds will suit the modern brown egg producer and you might have to work through several lines to find the one that actually works on your farm or in your poultry yard. You may even have to begin with but a few good specimens and launch a breeding up program.

To put a bit finer point on it, one of the best, most practical performing sets of birds we ever had here was a flock of about forty White Rocks producing hatching eggs for a small, regional hatchery. One of the poorest performing breeding groups we ever had here was also a small flock of White Rocks. The former, to give credit where it is due, were out of commercial hatchery stock, but they did have good size and exceptional vigor even in the presence of the large number of breeding males that are kept with a hatchery flock. The latter were truly massive and beautiful to behold, but had been bred solely for the showroom for generations and were lacking in vigor, fertility, and egg-laying ability.

The show-bred birds had an average body weight of two plus pounds over the far more productive group and were slow to nearly lethargic in their movement. They were slow to develop, also. The quest for size and truly elaborate coloring and feathering has led to some production ills. Many lines now have trouble breeding. Artificial insemination has become commonplace in many breeds and some now keep different breeding lines to produce males and females for exhibition. I recently had my interest dampened for starting a flock of Large Fowl Dark Cornish after learning how much artificial insemination is now employed in their propagation. A breed that cannot be advanced via natural and single matings has no place on a modern family farm.

White Plymouth Rock genetics remain perhaps the most accessible of all purebred poultry genetics, but time will still be needed to find the best fit for a particular farm. Breeding flocks and hatcheries based in northern climes sometimes offer more hardy birds due to the environments they face there. Smaller hatcheries with modest breed lists are good sources as they often offer top breeding from owned and controlled flocks of just those few breeds.

The Black Australorp was a bird selectively bred out of the Black Orpington breed specifically for exceptional egg laying performance. They were developed in Australia and hence the name Australorp. Many old timers believe that dark feathered, particularly black birds, will fare better and produce more than the lighter colored breeds in cold weather.

Be very mindful of size and type potential for egg laying when selecting Black Australorps. They aren't giants and taking them to excessively large size can be at odds with their role as egg producers. Good, black feathered birds have an appeal all of their own and this is one breed that also fits quite nicely in the multiuse category.

The Barred Plymouth Rock was an avian first love of mine and the quest for the clean, bright barred and clean yellow shanked and beaked birds I remember from my youth continues for many of my generation. The Barred Rock — not Dominecker — is a farm fowl deluxe from well back in the nineteenth century.

The Barred was the Rock breed for a great many people. Some truly legendary Barred Plymouth Rock flocks were developed and maintained well into the latter half of the twentieth century. They were a plain, tough chicken (in the good sense) and the breed has produced some of the most durable hens that I have ever seen. The vigor is still largely there and while some hatcheries are promoting their Barred Rocks on past glories, this breed has much to offer to those willing to put the time into them.

While many consider them just a tad too "plain vanilla," I have always been most partial to the white-feathered chicken breeds. The White Leghorn,

White Plymouth Rock, and White Wyandotte formed an economic troika that drove much of the development of modern poultry production.

While some may question just how many white-feathered chicken breeds are actually needed — most of them produced their bonafides back when the modest sized, working flock was the economic backbone of a great many American small farms and the families they supported. The Rosecomb Rhode Island White, the breed variety currently sanctioned by the American Poultry Association would be a good choice for those needing a moderate egg layer to produce in cold and harsh climates. The Single Comb variety was discussed in some detail previously and is a most utilitarian layer.

The Rhode Island Red and the New Hampshire are the classic, American red hen breeds although they do vary markedly in shading and color intensity. The Rhode Island Red came first in development and the New Hampshire then came about due to a desire for a red feathered bird with a bit more size for meat production and a durable nature to suit the family farms of that rugged New England state. The Rhode Island Reds by far have the broader gene pool, but neither should be confused with the hybridized "Production" or "Performance Reds."

A great deal has already been said here about the Buff Orpington, but for many, the Welsummer is a new name, although it is a quite venerable breed with some deep roots here and even more so abroad. It takes its name from the city of Welsum in the Netherlands. Welsummers and the Maran are the most commonly seen of the dark brown, egg-laying breeds.

The Welsummer is rather similar to the Light Brown Leghorn in coloring although with perhaps a bit more of a golden cast to the lighter colored areas. Some have encountered size issues with certain lines within the breed and for the past few years these birds have ridden along primarily on the demand for birds capable of laying distinctly colored eggs.

This is not to slight them as layers, but to rather point out the need now for some breeders to become very serious about making them better, more productive layers. They have the body type to be good layers and, as some of the novelty of the dark brown and heavily speckled eggs begins to wear away, their continuing success will depend upon how well and profitably they do lay.

They need to be continuously selectively bred for the darkest eggshell color possible and this certainly is a factor in their choice and propagation. However, this is not a factor that should prove detrimental in developing good, egg-laying flocks of this breed. Fortunately, many egg laying traits have a very high degree of genetic inheritability and by acquiring good foundation stock, selecting for size and vigor, and keeping replacement males from only the largest, most well shaped, and darkest brown egg produced flock, building this still relatively rare breed can continue apace.

All large chicken breeds, of course, lay eggs and differences in performance among birds and lines within a breed can often be even greater than the laying performance between some breeds. A line of Black Cochins, a breed not noted for its laying ability, could be developed that lays modestly well or even much above average for the breed. It will take much time and perhaps several false starts but it is possible. This line, however, could never be expected to lay as well as birds of even median performance for a breed like the White Leghorn or even some lines of White Rock.

A good producing, well-documented laying line of Cochins (of any of the very large breeds actually), of any color would have substantial value as long as they were of a good breed type. To a few this would be a challenge well worth undertaking, but for most, if the need is for brown eggs in goodly numbers then it is best to begin with a breed already proven and strong in that area.

Many who are now producing brown table eggs are doing so with one of the sex-link varieties with names like Comets or Cinnamon Queens. These are crossbreds and the crosses are made to produce baby chicks that can be sexed on sight at the time of hatching. Their strengths as egg layers come not from the cross, but the merits of their purebred parents. Heritage breeds may have been used in making these sex-linked crosses, but the birds cannot be considered as heritage material as they will not reproduce in kind.

Sex-link birds are a shortcut taken for economic ends and the crosses that produce them will never be exactly repeated. Many producers, who year-after-year purchase the same sex-link cross from the same hatchery, will undoubtedly relate how each new flock performs so differently from the others before. They are often rather high strung birds, have limited salvage value, and often produce eggs excessively large for their reduced body size. Many report losing large numbers of these birds to simple prolapse as they begin to lay.

Those needing the greatest number of eggs possible from a modest size area or to produce the most feed efficient egg-laying birds will best be served by the selection of birds from one of the white egg-laying breeds. This is the role for which they were developed and at which they excel!

LARGE WHITE EGG PRODUCTION BREED RECOMMENDATIONS

The White Leghorn has set the bar for egg production here and abroad, but terrible things have been done to this great old breed in the name of industrialized agriculture. Through the first half of the twentieth century the White Leghorn was as carefully bred and used as the finest and fastest of racehorses. With near artisanal breeding practices and time demanding testing practices such as trapnesting, the White Leghorn was bred to be a pearl-feathered gem that could be used by independent producers all across the nation.

The old literature and advertising is full of accounts from privately held Leghorn flocks that regularly averaged 280 and more eggs produced per hen per year. There were many hens that cracked the 300 eggs per hen per year barrier and some that produced 1,500 and more eggs in a lifetime. There are few hens now with a shot at even a second year in production let alone setting a four-figure lifetime record of egg production.

With the industrialization of egg production there no longer seemed to be the time for such careful and measured breeding practices and a rather arbitrary production type was settled upon for the confinement house layer. They would achieve desired levels of production with a small, hybridized bird that could be packed most snugly into cages or colony houses. They set upon a bird of small size that would produce a largish egg in proportion to their body size, that would lay heavily for a single season, and that could then be discarded as little more than a feathered husk with little or no value beyond mere landfill.

Such birds are but mere shades of what the White Leghorn once was. There are still some "big" White Leghorns about, but they have been bred almost exclusively for the show room and their laying performance has often suffered as a result. I have a good friend in Indiana, Matt John, who, like me, has spent many a year questing for the big, "English" White Leghorns that our grandparents kept so profitably.

Matt is now undertaking to combine various lines within the breed to recreate a White Leghorn with more than just good size, conformation and vigor. Into a line of birds with good size and real substance that he has in hand, he is breeding a strain noted for egg production. It is a commercial strain, but one still somewhat close to what the breed used to be. Through his efforts he hopes to create a laying and breeding flock of good size and on a par with the farm based Leghorn flocks of 75 to 80 years ago. This level of effort is what is needed with a great number of breeds now and is returning the birds to the form from which they were intended.

Actually, for most Americans the white-shelled egg is their traditional egg. It is virtually the only kind of egg that has been available in the retail trade for at least the last three generations of consumers. The white-shelled egg is nutritionally the same as the brown-shelled version and everything done to add value to the brown-shelled egg can be done with the white-shelled one, too.

The white-shelled eggs can be produced organically, cage free, and on range. With the breeds as they are currently available, it may actually be possible to produce more large and extra-large eggs from the white egg layers than the brown. The large, chalk white egg was the business card of the Leghorn breed and has been one of its selling points for decades.

Back in the day, the White Leghorn wasn't a massively large bird, but it had the frame and structure needed for a range bird. It also carried sufficient weight to assure that it retained some salvage value after their laying life is finished. This laying life could extend well beyond a second laying season — which is now double the norm for a caged layer.

At this time, I would not include the White Leghorn on a list of recommended heritage white egg layers. With only a handful of exceptions, the presently available laying lines have been bred for generations for a life in cages. I can't recommend this breed at present, even though this is the bird that shaped and defined modern poultry production. The thinking and planning behind this breeds early development went on to shape the management and care given to a great many other livestock species. But the role of trailbreaker has not been very kind to the White Leghorn breed.

My list would, though, include the Light Brown Leghorn, Ancona, Black Minorca, Buff Minorca, and then a possible selection from among one of the many colored Leghorn varieties. The Light Brown Leghorn was once propagated every bit as exacting and carefully as the White Leghorn.

Great effort and expense went into their early development. Many dollars were spent importing top birds from England and the Scandinavian countries. Sometimes, these are still referred to as Danish Brown Leghorns. A legendary Midwestern hatchery, now out of existence, built its reputation large upon a strain developed from Danish imports and some of that Danish breeding does still exist.

Light Brown Leghorns bred exclusively for exhibition will have sacrificed at least some productivity for the sake of size and intensity of color. The producer seeking to build a flock with this breed may have to shop widely and trial a number of lines to find the one that best fits his or her needs or they may have to build by combining birds from two or more different lines.

With the single comb Leghorns come concerns for comb and wattle damage in cold weather. Such injury is very seldom fatal, but it can disfigure birds and leave them infertile during the early part of the breeding season. Until the freeze damage heals, the bird may run a low-grade fever and the elevated body temperatures often kill sperm. To maintain good breed character you must select for a well-formed comb of good size in both males and females. Some old hands believe that a wide, well developed base to the single comb can better sustain the comb in very cold weather.

This color variety of the Leghorn breed is also bred in a rosecomb variety and this feature will certainly help a bird cope with cold climates. They are not widely seen and the best lines are probably to be found among individual breeders working to maintain solid breed character. In buying white egg genetics, I would opt first for birds or hatching eggs from more northern-based flocks. Birds from this region have been bred and must perform in the face of a more challenging climate.

The Ancona is a mottled, largely black with white tipping evenly distributed on the feathers that is also bred in single and rosecomb varieties. The rosecomb variety is not widely available. However there have been some small size issues with this breed. This was once a good sized, hard ranging breed that would forage well and had the mottled camouflage needed for additional predator protection.

Ancona have a reputation for producing proportionally large, chalk white eggs in relation to hen size, but this breed has languished for a great many years. I acquired a few not long ago and while the single comb gene pool for this breed is of some size — the birds had size, coloring and laying issues. Flock builders need to do some basic selection for size and vigor, beginning from the instant the chicks are removed from the hatcher tray.

The Black Minorca is perhaps the largest of the readily available white egg breeds and is certainly the largest of the Minorcas. This is a bird most striking in appearance and has been a favorite with both backyard breeders and exhibition breeders. It too is bred in both the single and rose comb types.

The Black Minorca is a breed still widely offered in the baby chick trade, but I suspect a number of hatcheries may be dipping into the same breeding flock pool or small handful of flocks to fill their egg needs. Some of these lines may have overall size concerns too.

With the Black Minorca, bird size and egg productivity must be kept in balance while building a flock true to the breed and its history of productivity. Producers should select birds not just for height, but real substance throughout the body. These birds are showy, but not avian wimps and some veteran showmen raise them using techniques that were once common in developing birds for the pit including penning males in individual, all-wire cages to develop.

The Minorca is also bred in buff and white varieties with the Buff Single Comb being the most commonly found after the black variety. The White Minorca is very rare. The buff variety will be smaller than the Black and a bit finer made as well.

The challenge with the Minorcas will be to find good birds with which to build a line. And while buff and black varieties will be available from a certain number of commercial hatcheries you are more apt to encounter smaller size issues there and the buff color is not the easiest color with which to work. It can fade or lose depth. Good Buff Minorcas will have a strong undercolor right down to the skin and I have been a part of discussions on this color that have literally gone on for an hour or more as producers share their thoughts on how to manage this color.

When you say "colored Leghorns," you have indeed said a mouthful. The 2007-2008 Directory of the Society for the Preservation of Poultry Antiquities lists the following Leghorn varieties; Single Comb and Rosecomb Dark Brown, Light Brown in both comb types, Buff in both combs, Black in both combs, Silver in both combs, Single Comb Red, Single Comb Black Tailed Red, Single Comb Columbian, Single Comb Golden Duckwing, and Single Comb Exchequer.

The Exchequer was sometimes called the "Scottish Leghorn" and was known as perhaps the largest of the Leghorn varieties. They have become more common in recent days, but color and size could be better and there is some real concern with leg color with this variety. Others exist in perhaps even smaller numbers and even mere mention of Black Tailed Red birds has been known to spark some real debate in some circles.

I have owned Dark Brown, Buff and Black Leghorn birds in the Single Comb variety and would like to have seen them all a bit bigger, but none lay on a par with the White and Light Brown varieties. Still, I think that all of the colors and patterns present in this breed offer elements of challenge and distinctiveness that should have more producers taking them up. They are birds that could do much to give white-shelled eggs a bit more pizzazz and an added hook with consumers.

The Leghorns are bred in still other colors and patterns here and abroad too. Much work, for example, is now being done with Large Fowl Leghorns in the Mille Fleur pattern. In a catalog I read not long ago I saw a "Barred" Leghorn advertised. In the accompanying text it was noted that it produced something of a tinted rather than a chalk white egg. This certainly raised a few red flags in the back of my mind. The Leghorn is "the white egg breed."

There are a number of other white egg producers including the Buttercups, Campines, Catalanas, Egyptian Fayoumis, Hamburgs, Barred Hollands, La Fleche, Lakenvelders, Redcaps, Norwegian Jaerhons, Appenzeller Spitzhaubens, and White Faced Black Spanish.

The Barred Holland were developed to be both a good layer of white eggs and to be a bird of some size. The Fayoumis may be the youngest to lay of all of the purebreds, although they produce rather small eggs throughout the whole of their lives. The Hamburgs were once termed "everyday layers" although they, too, produce a smaller egg. In Europe there has been some interest in the marketing of a smaller egg and there might be some market for it here in the United States among consumers concerned with portion control for health or diet reasons.

Some of the breeds just mentioned have gone on to be bred more for their appearance than their egg-laying ability. The White Faced Black Spanish may be one of the most distinctive appearing of all poultry breeds. Yet I have a January 1919 issue of *Farm Journal* that touts the egg laying value of such white egg-laying breeds as the La Fleche and Houdan from France.

MULTIUSE BREED RECOMMENDATIONS

A large fowl category that has garnered a lot of attention in recent years is what many termed the "dual purpose" group. This is a bit of a manufactured grouping and I think that the term "dual purpose" may be a bit misleading.

The birds in this group do not combine the best traits of the egg breeds and the meat breeds. By its very design a good layer will not be a high yielding meat bird and a good meat bird does not have the body type of a good layer. The birds in this group are, at best a compromise that combines modest performance in both traits. The bird that lays eggs like a good White Leghorn and that also grows and yields like a Cornish X bird would be a "Super Chicken" and not simply a dual-purpose fowl.

More and more one hears the term "multiuse" to describe these breeds and I think it is the better and more descriptive choice for these birds. If you want eggs in goodly number and poultry meat rapidly produced — opt for two different breeds bred more specifically for those separate purposes. The birds I list here are utilitarian in their nature and will produce eggs in fair number from simpler facilities while having some good dressing qualities.

You will note some overlap in this grouping and lists I have given elsewhere in this text. Selective breeding can move some breeds more toward one set of traits or another and within some large fowl breeds you will find lines that lay better or grow and yield better than others.

Those most interested in multiuse fowl seem to be producers wanting birds that will perform in simple facilities and meet family needs. Valued are birds with the hardiness and vigor needed to forage and that can be maintained without a lot of artifice.

On this multiuse list I would place Barred and White Plymouth Rocks, Delawares, New Hampshires, White and Columbian Wyandottes, Buff Rocks, and Buff Orpingtons. With this list I know I certainly won't please everyone and many more are the subject of gentlemanly discussions about what breeds to add or delete from this list.

My choices here do reflect my Midwestern upbringing and practical bent and all but one have the yellow skin and shanks so favored in dressed chickens in this country. The Barred and White Rocks were discussed extensively earlier in this chapter, but in this particular instance I would rank the Barred over the White variety.

The Delaware is a relative newcomer to poultry circles, not fully arriving on the scene until after World War II. It just barely qualifies as a heritage breed under the timeline now set down by the American Poultry Association. The Association now declares heritage breeds as those in existence up to 1950 and no later. Yet this breed was one of the key ones in the development of the modern broiler trade.

In crosses with the New Hampshire, the Delaware was used to develop the Indian River broiler, an early, hybrid broiler variety. The Delaware is not a true Columbian patterned bird, but rather the dark feathering in the hackle and elsewhere carries a form of barring. In crosses with red birds and birds that carry what some call the "gold" gene for coloring, it manifests in much the same way as the barred gene of the Barred Rock does when used in such crosses.

When bred pure, the Delaware can be selectively bred and developed to produce large, light brown eggs in fair numbers or birds of a fairly good meat type. One noted poultry seedstock producer, candidly reported a few years ago, that he had perhaps gone too far in selective breeding for large sized eggs with this breed. His line began to encounter prolapses and other problems associated with excessively large eggs.

Good Delawares are still not as common as advertised sources for the breed might lead you to believe, but this bird comes mighty close to epitomizing what "multiuse" should really mean. Also, it is a very pretty chicken.

Of the red breeds, the New Hampshire, in its type, may be the most utilitarian. Until quite recently there was a quite precipitous decline in New Hampshire numbers and some real concern was being expressed about its future and the genetic purity of the remaining populations.

When a claim is made for being a part of a particular line within this breed it usually traces back to just one of two lines once dominant in the breed. New Hampshires were used in more than one early broiler cross and it has the frame size and overall outline to be a fairly good meat bird — in the earlier tradition of classic frying and roasting chickens.

The White and Columbian Wyandottes are choices that right now are skewing more toward the poultry meat role of multiuse fowl. These are personal favorites of mine, but both need a lot of work in the area of egg production. They are of a good size and present with a frame and muscling pattern conducive to poultry meat production using simple housing.

These Wyandottes have feathering and rosecombs that make them a good cold tolerant bird, too. Seventy-five years ago this breed did have some good laying flocks into the colder regions of this country and were producing eggs on a per hen level in the same range as some good flocks of Rocks.

The Columbians tends to be a bit smaller than the White Wyandottes, but are not nearly as widely available. The White Wyandotte perhaps represents less of a challenge to the breeder than the even more common laced varieties of the breed. In England White Wyandottes are now being bred in both exhibition and for more utilitarian purposes. The hens of the latter type are often bred with Rhode Island Red males to create very practical appearing hybrid layers with some scale and added durability.

The Buff Orpington has been well discussed here, but is a good winter hardy bird that was first developed for the meat trade in Great Britain in the area from which it takes its name.

The Buff Plymouth Rock may be the smallest of that variety, but could be the "easy keeper" that many are looking for when they contemplate a multiuse bird. Also, this is one of the varieties that is more apt to go broody and this is a trait many seem to want in a small, multiuse poultry flock.

Buff Plymouth Rocks are available fairly widely, but, again, single flocks may be supplying multiple hatcheries as this variety has much reduced status from many other Plymouth Rock varieties. With this variety, as with many others, begin with at least twenty-five as hatched chicks. Then cull hard for size, vigor and good breed character. Also accept that you may wind up with but a handful of keepers in the early years of flock building. They will make a postcard pretty flock and that eye appeal can and should be incorporated into the marketing of the meat and eggs that they produce.

I know I have sparked some controversy with this particular list and can close my eyes, lean back in my chair, and hear some of you saying, "Come on, Klober, where are my Dominiques or Javas or Langshans or so many more?"

To answer that, at least in part, please note that my Rhode Island Reds and Whites, good old breeds both, didn't make my list either! I meant the list to be one of work horse breeds, fairly easy to access, with fair egg laying and dressing qualities, with a certain amount of eye appeal, and with birds that would not present too great a challenge to beginning producers. Producers are not going to pop major show winners out of a hatchery run box of little Barred Rocks. But in this listing there is a solid group of birds where good breed type is strongly tied to the roles they were meant to play and for which they are again being sought.

My grandmother's favorite chickens to fry were young Leghorn cockerels hatched fairly early in the year and harvested in the cooler days of the fall. Not long ago I set for a while with a good friend and poultryman, Allan Burgin as he shared memories of his mother's flock of hard working White Cochins. They were a bit seasonal in their laying patterns, but they hatched and raised their own replacements. Surplus cockerels of that strain could be grown to be fair meat birds, and there was always a good demand for any extra of these stately birds that she cared to sell.

Just about every breed could have multiple uses if you are reasonable in your expectations for them. I have one friend very enamored of the Partridge Plymouth Rock, another more than happy with his French Cuckoo Marans, and one enjoying the challenge of a small flock of Blue Orpingtons that win at nearby shows and supply his family with eggs and fryers.

"Multiuse" is a term that denotes a bird of some practical merit although not one that is a top performer in either meat or egg production. Every breed must lay some eggs, but there are those like the Cornish that lay a relatively few eggs and those like the Fayoumis that lay rather small eggs. All chickens will also fry, it's just that some breeds mature at smaller sizes and others, while large as adults, may mature rather slowly.

You cannot put everything good about the whole chicken tribe in but one single breed. For very best results a producer needing both poultry meat and eggs will be best served by keeping two separate flocks. The larger, laying flock should be of one of the breeds known for its egg laying strengths. A smaller flock of one of the meat breeds can produce a good many eggs to be hatched each year to produce chicks to be grown out as broilers and roasters.

MEAT BREED RECOMMENDATIONS

A huge quest has been underway for sometime to develop purebred alternatives to the Cornish X broiler. Not ounce for ounce replacements, but an alternative with good growth and yield and which can be bred and propagated on family farms all across the land.

Well into the twentieth century chicken was truly a premium food item, valued on a par with prime beef. Think how often politicians mentioned chicken as a symbol of the boom times they hoped to bring to the land. Agribusiness did lure greater numbers and cheaper chickens to this country or, at least, what they envisioned chickens should be.

Grandma's chicken looked and tasted differently than these new imports and for a number of reasons. Yesterday's poultry wasn't bred to have a monstrously disproportionate breast or to grow insanely fast on a high concentrate ration. And, it really did taste differently — like chicken should.

Grandma's birds grew slower, ate a more varied diet, ranged about and got more exercise, and were, generally, purebred birds developed and hatched in her region of the country. When she and Grandpa dined on chicken they were, indeed, eating locally. A Cornish X bird is now harvested at between five and eight weeks of age and has lived almost entirely on a crumbles or meal diet. Most purebreds now take eleven to fourteen weeks to reach a good harvest weight, but in so doing the meat they produce is imparted a more natural texture and a fuller flavor. Serving sized pieces are apt to be smaller, however.

Many studies have now shown that pork and beef from different pure breeds has a unique flavor and other good eating qualities. Some very preliminary taste tests have been done with poultry meat from a handful of the heritage breeds, but the data thus far gathered is little more than anecdotal. For example, the Dominique breed is said to produce meat with a taste and in serving sizes quite reminiscent of pheasant. So there's good potential starting to show for the future of unique tastes in heritage breeds too.

Other breeds have been featured in taste tests done by major chefs and food writers. The results report that the distinctive nature of the birds comes through even when prepared in a number of different, truly seasonal poultry dishes and preparation styles. Much more testing is needed to verify these perceived differences and this data then needs to be used to build a solid position in the marketplace for these purebred birds.

Right now, pop into a meat bird show at any county or state fair and you will find it totally dominated by Cornish X birds. Generally, they will be exhibited as meat pens of three cockerels and weighing four to six pounds each. At a very few, you might find a judge willing to entertain the idea that a market bird can be something other than a Cornish X bird. At these events a few alternatives have found their place and at our own county fair a couple of years ago a young lady won a blue ribbon with a pen of well grown, young White Plymouth Rock cockerels.

The winning birds were about twelve weeks old and there lies the greatest concern about developing alternatives to the super fast growing, Cornish X broiler. They are going to be on feed and on the farm for an extra four to six weeks. Testing shows that more of them will survive to reach harvest, few will have to be discarded due to leg problems and breast blistering, and they will forage for more of their sustenance. On the other hand, with the purebreds, there has been little or no selective breeding for growth and yield in well over half of a century.

With time and attention to breeding details, many believe that purebreds can be put to a harvestable weight of four to five pounds by ten to eleven weeks of age and still retain their nature as real chickens and not feathered Frankenfood. Feed efficiencies can also be made better, too.

The list of breeds that I would offer as candidates to produce a good poultry meat flock comes with one proviso. That proviso is that producers take them up both for what they are now and what they can become using a well thought out breeding plan. Breeds like the Black Giant (the "Black" comes not from the bird's color but is the family name of the brothers who were the ones involved in the early development of the breed), Java, New Hampshire, and others were the early building blocks of the poultry meat industry and many may yet have a role to play in the future.

The New Hampshire would go on my list today, as would the Delaware, White Plymouth Rock, White Wyandotte, and the Rosecomb Rhode Island White.

These birds will dress out with a yellow skin and shanks, will take a good finish, will produce carcass pieces of good size, and all but the last breed named are quite readily available. You may have to work through the offerings of a number of hatcheries and breeders to find the right combination of growth and vigor to fix in a producer's farm/market centric line.

You won't want to draw from show bird lines that have been bred for too massive size nor lines that have been bred extensively for egg production. Go back to some of the earlier American Poultry Association Standards of Perfection for descriptions to truly correct the productive type and size for these breeds. Use these as a guide for selecting foundation birds and building a breeding flock.

The White Wyandotte has long struck me as a bird with real potential for development as an alternative meat bird. It has the size, good muscling, breed traits (including rosecomb for harsh weather), and history to be a most natural fit for the growing interest in localized agricultural production. It won't be done overnight and a third or fourth start may have to be made in finding breeding pieces, but here is a bird with a whole slew of pluses.

The Rosecomb Rhode Island Whites I have had in hand have been very impressive as they move toward maturity, but they are a bit slow to grow and can go through a real "gangly" period at around eight to twelve weeks of age. Here, too, is a bird with a lot of potential in the "local trade" and that can be made into an ever better meat bird.

The meat bird trade is not just limited to broilers for frying and roasting, either. The growing numbers of ethnic communities in this country have given rise to a demand for black and/or red feathered birds of several weights and ages as customers seek out birds similar to those they remember from back home. Many hatcheries now offer black and/or red broiler chicks.

The "blacks" may be purebred Australorps or Giants or a cross of the two. Many of these markets are very sensitive to rather small details and will not buy a black bird with a forked comb such as is seen on the black La Fleche. Or they may prefer males to females or want different colors for special events. Many are adamantly against white-feathered birds, too. Even a crossbred of the first two black breeds will not be the most efficient of meat birds, but much could be made of them in a relatively few generations of selective breeding. Top show specimens of the Black Giant now may take a whole year to fully develop and fill out.

I think a faster growing Australorp is quite doable and would respond in as few as three to five generations. The Black Langshan offers some interesting possibilities, too, but as more of a niche bird. One of the best choices for a red broiler would appear to be the New Hampshire. They grow reasonably well, but the males do start taking on strong secondary sexual characteristics as they approach a harvestable size. This is true of most of the young purebred males as they will take on a rooster look as they grow rather than the asexual appearance of a Cornish X bird.

Two other rather interesting alternative choices for use as meat birds are the Buckeye and the purebred Cornish large fowl. The Buckeye is close to the Rhode Island Red in its deep, mahogany red coloring. It has a small, tight comb and the males are measurably larger than the females. It is an old breed, now relatively rare, but it has been the focus of some well-organized preservation and promotion work. It too is a bird with a lot of historical and regional appeal. The gene pool still needs some broadening and selection for improved growth would be in order, but this is a possible meat bird with a lot of potential.

The Cornish, as it was used extensively to develop the Cornish X bird, would seem to be a natural for alternative meat bird producers. However, they are not the easiest of breeds with which to work. Their challenges include fertility problems, quite limited numbers, upright stance from early development as fighting stock, tendency to lay nearly round eggs, modest egg laying numbers, and their tendency to be more seasonal breeders and layers. Additionally, artificial insemination is being used to develop and produce the extra large Cornish X birds now being exhibited.

In this country they are bred in White, Dark and Laced Red cornish varieties. The White cornish is very rare and my research has turned up only a handful of breeders for this color variety. The Laced Red is a most colorful bird and one that has been bred nearly exclusively for exhibition purposes. Those I have owned have had both vigor and fertility issues, but this is often true of birds bred for size and color at the expense of all other traits.

The Dark Cornish is by far the most commonly seen of the Cornish varieties, but even with this variety I have noted some real ups and downs in size and vigor. Many are quick to say Darks are breeder or hatchery birds, but the hatchery birds do naturally mate although they may be coming from a rather limited number of small flocks. At the 2008 Acres U.S.A. Conference in St. Louis I met a young Amish man that had grown out a modest set of Dark Cornish chicks as an alternative to the Cornish X birds that were most widely available in his area.

Dark Cornish do grow rather slowly, but he could begin harvesting some of these birds at the smaller size from which classic "game hens" are known. His birds would be the real deal rather than Cornish X pullets harvested at a very few weeks of age and lacking in flavor and character. When dressed, the older birds did present with larger breasts and the grower reported that they found much favor with those wanting a true, high end table bird.

The purchase of a small group of as hatched Dark Cornish chicks might be a good "experiment" for those seeking a new or different meat bird. A breeding flock of Cornish large fowl will be no simple undertaking, but they are a breed that many find intriguing. That they continue to appear in so many of the commercial hatchery catalogs indicates the relatively high degree of interest in them.

HISTORICALLY IMPORTANT BREED RECOMMENDATIONS WORTH INVESTIGATING

Lastly, I would like to propose a small handful of breeds that fit no easy grouping, but that do have some practical roots to them. These have been historically important and still have economic and practical roles to play. Into this diverse list I would place the Maran, the Polish, the Naked Neck, the Silver Laced Wyandotte, the Black Langshan, and the Light Brahma.

The Maran may lay the darkest of the dark brown eggs. They are seen in two varieties; the English (or clean legged) and the French (or feather legged type). The French are actually partridge legged with modest feathering down the outside of each leg. Many favor the French for their vigor and a bit of a size advantage.

Nearly all the English Marans trace back to a large, intensely bred flock, founded on French genetics brought to England prior to World War II. The breed was developed to hustle and forage in the farmyards of one of the more damp and chilly regions of France. While other color varieties are being cultivated in both varieties in this country, they are most often seen in the cuckoo pattern, a pattern of rather uneven and muted barring.

I favor the French variety myself and this does seem to be growing in numbers as more poultry raisers gain experience with the breed. The good ones are still primarily available only from individual breeders and should be bought from those selecting extensively for vigor and the darkest, most intensive eggshell color.

I know that by including the Polish breed above many must have said, "What?" Isn't this the bird with the outrageous crest and the often flamboyant color patterns. Yes, this is correct, but this is a bird that was developed for small farms in cold climes that can lay a large, chalk white egg. The Polish breed has also been bred for quite good size in many of its lines, and it draws a lot of attention with potential buyers whenever and wherever it is displayed. I have owned White Crested Black, Buff Laced, and Silver Laced Polish varieties and enjoyed working with them. The Polish is bred in a near two-dozen colors and patterns and in bearded and non-bearded varieties. The first fancy bird I owned as a kid was a White Crested Black Polish rooster.

Beginners should seek one of the simpler colored varieties and a few, non American Poultry Association sanctioned varieties are bred in but a single color. I have seen Whites and Blacks and have been reading about a Red Polish variety. Polish breeders always seem to be working on a new color or pattern.

Do select for birds that are generations deep in the preferred color or pattern. Likewise, select carefully for good size and vigor. Due to the crest, the Polish should probably not be closely penned with others — especially more aggressive breeds as they have a very limited field of vision.

Producers really have to be a student of this breed. Its unique appearance has given it great recognition. It is far from the equal of the Leghorn or Ancona, but for someone seeking a truly unique white egg layer and on which they can place a rather personal spin, a small flock of these might be grown and developed into a modest, side venture. They certainly appeal to today's suburban and backyard poultry keepers seeking more unusual birds in small numbers from more localized sources.

Now, about the Naked Neck breed, the name "Turken" is really something of a misnomer. Its early development was as a meat bird and it still has the size and good growth, and it can produce a quite good-sized egg. Recently it has been taken in hand by some dedicated breeders and is being bred in a number of true breeding colors with red and black being perhaps the most common. Also, a friend has developed a very good barred line in this breed.

Naked Neck is a breed that is not going to appeal to everyone, but it is a very distinctive appearing bird with good response to most practical applications. Genetically, there is far more to this breed than a naked neck and breeders are getting very serious about making this a real working breed again.

The Silver Laced is the Wyandotte variety most of us have probably seen, but I expect that very few have seen real good examples of this breed variety. This was the first variety of the Wyandotte breed developed and some of the others occurred as sports coming out of it. It is a challenging pattern to breed for and some of the lines in commerce lack good Wyandotte type. They are instead angular representatives of what is supposed to be "the bird of curves." There have also been some issues with bird and egg size and egg productivity.

The Silver Laced is a most striking bird, cold hardy, and the variety most think of when Wyandottes are mentioned. Genetics are accessible and they may be one of the breeds suited to small flock producers looking for a breed with both challenges and practical applications.

I have long mulled over including the Black Langshan in this list, as its long history was a factor. It now has new breeders, produces a rather dark brown shelled egg, has size, and a well-muscled breast. It is a tall, distinctively shaped breed with moderately feathered legs. It is bred in black, white and blue with the black variety, by far, the most commonly raised. What will be made of this breed remains to be seen, but it has a following and may be the most athletic and vigorous of the large, feather legged birds. And there is that wonderful black color.

The Light Brahma is a feather legged bird, tall and a bit slow to develop. However, early in the twentieth century, eggs of a Missouri developed strain of this breed were sent to Australia where they were hatched and the chicks grown out to defeat some of the best locally bred Australorps developed there in egg laying competitions.

There is tremendous eye appeal with this breed and the good ones just seem to shine when raised on grass and in sunlight. I have memories of small flocks of this breed on Midwestern family farms going back fifty years and they still appear in many commercial hatchery catalogs, always a yardstick of demand. There can be challenges with getting these birds bred if size or feathering is carried to too great of an extreme.

With its size and unique head structure, Brahma draw lots of attention and are a breed many remember from an earlier time. I looked long and hard at a set of Brahma pullets last spring at our local Spring Poultry Expo. They even caught the eye of Phyllis, my most practical wife.

Indeed, there are breeds that still can sway me and draw my eye. There are birds beyond what are in my poultry yard here that I am open to and this is, I think, true of nearly everyone in the poultry fraternity. Something comes along every once in awhile that you just can't turn down, whether due to its rarity or the quality of the stock being made available. You can't own them all, but few are the poultry folk without room for at least one more, special breed.

The breed choices set down here are just that — mine. I can give what I believe are some most practical arguments for them, but others have their favorites and their reasons for them too. However, everyone considering breeds should take the time to learn the poultry keeping history of their region, the breeds that worked well and were valued there, and stay plugged into what is going on in the poultry scene right now.

That Queen Victoria kept Buff Cochins, no longer helps the demand for these birds, but a Salmonella scare highlighted on the evening news does step up the demand for chickens like these that your grandparents used to keep. These historic birds included those that were humanely kept and in a content way, were practical and productive, produced eggs and meat that tasted like eggs and chicken should, and that were safe and left consumers with a "feel good" experience when they bought them.

Here, found in the wants and needs of an informed consuming public, combined with the need for a truly sustainable agriculture, is the future of the heritage breeds. Your part in preserving heritage breeds may be in keeping a flock of five, ten, fifty or five hundred or in keeping a poultry yard with one breed or ten. Whatever your flock size, the key is to give these traditional breeds an opportunity to show their best qualities on your farm.

CHAPTER 11

HERITAGE POULTRY BREED PHOTOS

Silver Ameraucana cock.

Image courtesy of Will Morrow.

Silver Ameraucana hens.

Image courtesy of Will Morrow.

Ancona hens.

Image courtesy of Yellow House Farm.

Ancona rooster and hens.

Image courtesy of Yellow House Farm.

Barred Plymouth Rock cock.
Image courtesy of Elaine Belanger.

Barred Plymouth Rock cock.
Image courtesy of Elaine Belanger.

Hamburg cock.
Image courtesy of Elaine Belanger.

Hamburg hen.
Image courtesy of Elaine Belanger.

Buckeye cock.

Image courtesy of Laura Haggarty.

Buckeye hen.

Image courtesy of Laura Haggarty.

Buckeye flock in the snow.

Image courtesy of Laura Haggarty.

Buckeye flock.

Image courtesy of Laura Haggarty.

The Buckeye's pea comb.

Image courtesy of Laura Haggarty.

Buckeye flock.

Image courtesy of Laura Haggarty.

Buckeye hens and rooster.

Image courtesy of Laura Haggarty.

Buckeye flock.

Image courtesy of Laura Haggarty.

Delaware cock.

Image courtesy of Will Morrow.

Delaware cock and pullet.

Image courtesy of Will Morrow.

16-week-old Delaware pullet.

Image courtesy of Will Morrow.

A top-performing Delaware cockerel.

Image courtesy of Will Morrow.

Dominique pullet.

Image courtesy of Bryan K. Oliver.

Dominique cockerel.

Image courtesy of Bryan K. Oliver.

Dominique cockerel and pullet.

Image courtesy of Bryan K. Oliver.

Dominique rooster.

Image courtesy of Bryan K. Oliver.

Dorking hen.

Image courtesy of Yellow House Farm.

Dorking rooster.

Image courtesy of Yellow House Farm.

Dorking cock.

Image courtesy of Yellow House Farm.

Dorking hen.

Image courtesy of Yellow House Farm.

Java hen.

Image courtesy of Mike Dougherty.

Java cock.

Image courtesy of Mike Dougherty.

Java flock. Image courtesy of Mike Dougherty.

Blue Maran hens.

Image courtesy of Kathleen LaDue.

Splash Maran cockerel.

Image courtesy of Will Morrow.

Black Maran cockerel.

Image courtesy of Will Morrow.

Cuckoo Maran cockerel.

Image courtesy of Will Morrow.

Black Copper Maran cock and Blue Copper hen.

Image courtesy of Will Morrow.

Splash Maran pullet.

Image courtesy of Will Morrow.

Blue Copper Maran trio. Image courtesy of Will Morrow.

Wyandotte cock.

Image courtesy of Elaine Belanger.

Wyandotte cock.

Image courtesy of Elaine Belanger.

Wyandotte cock.

Image courtesy of Elaine Belanger.

Wyandotte cock.

Image courtesy of Elaine Belanger.

CHAPTER 12

POULTRY
RESOURCES

(0) Beak
(1) Comb
(2) Face
(3) Wattles
(4) Ear-lobe
(5) Hackle
(6) Breast
(7) Back
(8) Saddle
(9) Saddle feathers

(10) Sickles
(11) Lesser sickles
(12) Tail-coverts
(13) Main tail feathers
(14) Wing-bow
(15) Wing-coverts, forming wing-bar
(16) Secondaries, wing-bay

(17) Primaries, or flight feathers
(18) Flight-coverts
(19) Point of breast bone
(20) Body & fluff
(21) Thigh
(22) Knee-joints
(23) Shanks
(24) Spur
(25) Toes or claws

GLOSSARY

Air Cell — The air-filled pocket between the white and shell that is usually at the large end of the egg.

American Standard of Perfection — The breed standards book recognized and published by the American Poultry Association.

Anthelmintic — to destroy or expel worms, usually in the intestines.

Artisanal — a product made by hand by a skilled craftsman.

As Hatched — These are newly hatched chicks that have not been sexed. Also called unsexed or straight run chicks.

Bantam (Banty) — Small-size poultry that have mature weights of less than two pounds and most often are bred as a treat for the eyes and a challenge for the breeder's arts. Bantams are usually 20-25% smaller than their full-sized counterparts. Most are smaller versions of large breeds but some such as Silkies are true bantams.

Broilers — Young poultry raised for their tender meat. Also known as fryers.

Brood — A female hens "mothering" of a clutch of eggs and chicks. See broody. A collective term for a clutch of chicks. Also called "setting or sitting."

Brooder — An enclosure, often heated, to protect newly hatched chicks.

Broody — After all of a clutch is laid, hens now remain on the nest keeping the eggs warm for 21 days and may become overly protective. Often called "clucking."

Candle — To hold an egg up to a bright light to inspect for cracks.

Capon — a castrated male chicken meant to be fed to a heavier market weight.

Chicken Tractors — Moveable range units that can be repositioned daily (or more frequently) around the pasture to give poultry new foraging areas.

Clean legged — A poultry breed with no feathers on their shanks (legs).

Clear Eggs — Eggs that have not been fertilized and show no embryonic development..

Clutch — A batch of eggs hatched together in one hens laying cycle. Also called "setting".

Coccidiosis (coccidiostat) — A parasite disease (and chemical used for the treatment thereof) found in the intestinal tract that can interfere with food digestion and nutrient absorption. Severe cases can result in dehydration, blood loss and even death.

Cockerel — A young, male chicken under one year of age.

Comb — The fleshy, often red structure on the top of a chicken's head. Often more pronounced on males.

Coop — A house or cage to protect and corral chickens.

Crest — The tuft of feathers on the heads of some poultry breeds such as Silkies and Houdans.

Cull — To remove a non-productive chicken from the flock or a term to define the inferior chicken.

Dam — A female parent hen.

Debeak — To remove a portion of the top beak to prevent cannibalism of others or injury to the bird itself.

Egg Tooth — The horn-like top of a chick's upper beak that helps break through the eggshell.

Feather Picking — Chickens under stress may pluck the feathers of another bird or their own. This pecking can become obsessive once blood is raised and can lead to near cannibalism unless an intervention takes place.

Flock — The entire collection of poultry living in one coop.

Forced-Air Incubator — An incubator that a fan to circulate warm air.

Frankenfood — The term for genetically modified food.

Free range — Allowing chickens free access to a yard or pasture as they please.

Fryers — See broilers.

Gamecock — A male bird with the physical and behavioral traits necessary from cockfighting. This includes being more territorial, having less prominent wattles and combs and possessing hard feathering. Often describes a type of bird but not a specific breed.

Geegaws — A decorative bauble or trinket.

Genotype — The full set of hereditary gene information, including mainly of the invisible traits.

Grit — Sand and small coarse material used by the chicken in their gizzard to grind up grain and plant fiber.

Hen — A mature, female chicken.

Heterosis — This is also known as hybrid vigor or outbreeding enhancement and describes the increased strength found in hybrids. Breeding for heterosis plays to the possibility of obtaining a genetically superior individual by combining the virtues from its parents.

Hybrid — The offspring of a hen and rooster of different breeds.

Incubate — To provide warmth and protection for hatching fertile eggs.

Incubation Period — About 21 days, the time that an egg takes to hatch.

Incubator — An enclosed device for hatching fertile eggs. It may have a fan and rotating racks.

Intensity of Lay — The output of eggs that one hen lays during a certain period.

In Toto — This means in its entirety or completely.

Keel — the keel is a bone on the sternum that holds the breast muscles. The keel bone typically should be straight for quality-bred poultry.

Label Rouge — A strict production method originating in France that emphasizes specific standards that respects animal welfare and protects the environment.

Linebred, Linebreeding, or Inbreeding — These are all terms for the breeding of closely related animals such as distant cousins who share a related ancestor. This is done to fix desirable traits without facing the greater risk of undesirable traits appearing with even closer related breeding.

Litter — The straw, wood shavings, shredded paper and other materials that are used on the floor of the chicken coop to absorb moisture and manure.

Loose-Housed — A term to describe birds that are technically housed for twenty-four hours a day but are not caged. They will generally have two to four feet of floor area per bird and may or may not be given range or lot access. The birds are at least able to move about and turn freely.

Molt — The once a year shedding and regrowth of a chicken's feathers.

Nick, Nicking — Having extra success in the way that bloodlines perform when mated. When the result is an exceptional offspring, better than its parents, it is said that they nicked well.

Outcrosses — These are matings of birds of different strains from within the same breed.

Perch — The place where chickens rest at night. Also called "roost."

Phenotype — The visible, both physical and behavioral hereditary traits.

Pip — The hole that a chick makes in its shell just before hatching.

Pit Game — Those birds are raised for cockfighting. Pit Game poultry are not a pure breed but chickens bred specifically for this purpose.

Poult — Any young fowl.

Poultry — Chickens and other birds that have been domesticated for eggs, meat or pets.

Preponent — This is the ability of an individual or strain to transmit its characteristics to offspring because of homozygosity through numerous dominant genes.

Pullet — A young female chicken under one year of age.

Purebred — The offspring of a hen and rooster that are of the same breed. Also called "straightbred."

Rolling Mating — The breeding plan to take the best pullets produced and breed them to the best males of the first (parent) generation.

Rooster — A male chicken.

Seedstock — A pool of new gene supply and the source of new individuals.

Sexed — Newly hatched chicks that have been sorted into males and females.

Sex-Linked — These are poultry bred first and foremost for sight sexing at hatching. These are the product of crosses involving the gold and silver color factors or barring gene used to create baby pullets and cockerels that will differ from each other in down color or pattern.

Spent Hen — This is a breeder or layer hen that is not performing up to a desired level.

Spraddled Legs — a problem with mainly young poultry when their feet slip apart. They don't have the strength to bring them back close together again and they stop walking.

Started Pullets — Young females that are nearly old enough to lay their first egg.

Starter — The first feed rations that newly hatched chicks are fed. Also may be called "crumbles."

Terminal Cross — A breeding term when the offspring are not going to be used for future breeding programs or genetic improvement. Usually these offspring are market birds.

Trapnesting — A technique used by poultry breeders where a special trap door is released after a hen enters a nest box, confining her while she lays her egg so it can be properly marked for identification purposes.

Variety — A division of a breed into different colors, comb style, beards or leg feathering.

Vent — The exterior opening with two channels through which chickens lay their eggs or excrete manure.

Wattles — Two fleshy structures hanging down from a chickens throat or chin. They are used to cool the blood that circulated through them on hot days.

POULTRY BREED CLASSES

LARGE BREED CLASSIFICATION AS IDENTIFIED BY THE AMERICAN POULTRY ASSOCIATION

American Class — Plymouth Rocks, Rhode Island Reds & Whites, New Hampshires, Jersey Giants, Dominiques, Wyandottes, Javas, Buckeyes, Chanteclers, Lamonas, Hollands, and Delawares.

Asiatic Class — Brahmas, Cochins, and Lanqshans.

English Class — Dorkings, Red Caps, Cornish, Orpingtons, Sussex, and Australorps.

Mediterranean Class — Leghorns, Minorcas, Spanish, Andalusians, Anconas, Catalanas, and Sicilian Buffercups.

Continental Class — Hamburgs, Campines, Lakenvelders, Barnevelders, Welsummers, Bearded & Non Bearded Polish, Houdans, Faverollis, Crevecoeurs, and La Fleche.

Other Standard Breeds — Modern Game, Old English Games, Sumatras, Malays, Cubalayas, Phoenix, Yokohamas, Aseels, Shomos, Sultans, Frizzles, Naked Necks, Araucanas, and Ameraucanas.

BANTAMS CLASSIFICATION

Single Comb Clean Legged — Anconas, Andalusians, Australorps, Campines, Catalanas, Delawares, Dorkings, Dutch, Frizzles, Hollands, Japanese, Javas, Jersey Giants, Lakenvelders, Lamonas, Leghorns, Monorcas, Naked Necks, New Hampshires, Orpingtons, Phoenix, Plymouth Rocks, Rhode Island Reds, Spanish, and Sussex.

Rosecomb Clean Legged — Anconas, Antwerp Belgians (d'Anvers), Dominiques, Dorkings, Hamburgs, Leghorns, Minorcas, Redcaps, Rhode Island Reds, Rhode Island Whites, Sebrights, and Wyandottes.

All other Clean Legged — Ameraucana, Araucana, Buckeye, Chanteclers, Cornish, Crevecoeurs, Cubalayas, Houdans, La Fleche, Malays, Polish, Shamos, Sicilian Buttercups, Sumatras, and Yokohamas.

Feather Legged — Booted, Brahmas, Cochins, d'Uccle Faveroles, Frizzles, Langshans, Silkies, and Sultans.

POULTRY FEATHER COLORATION DESCRIPTIONS

Barred — white feathering with black or a very dark color as "bars".

Birchen — Silver to white head, white neck and upper breast with a slender black stripe down the middle transitioning to black. Lower body black.

Columbian — white or silver white body with some black feathers with white lacing on the hackle and tail.

Crele — Males have pale straw colored feathers barred orange red and grayish white. Females have pale gold feathers with grayish brown barring. Rest of the body is dark gray with some barring.

Cuckoo — Bluish white feathers with irregular light and dark bars.

Exchequer — white and black colors mostly evenly distributed.

Henny — males lacking sickle feathers that look similar to females.

Laced (single or double) — single ring color around the outer edge of the feather with a different color filling the interior. Double has two rows of a color with another between.

Mille-Fleur (Millies) — A French word meaning a thousand flowers, a background made up of many small flowers or plants. In poultry, the background color is usually mahogany or orange with white or greenish black accents.

Mottled (Splash) — black with white v-shaped splashes on every two up to every five feathers.

Penciled — narrow bands of a different color across the width of the feather.

Porcelain — a background color of straw with blue barring with a white spangle highlights.

Pyle — Solid colors of orange, gold, salmon, white and red shading.

Spangled — silver color with sharply contrasting black at the end of each feather.

The American Livestock Breeds Conservancy issues an undated Priority Watch List for large fowl breeds in the United States. This list includes breeds that have had an established and continuously breeding population in the United States since 1925 or have been imported or developed since that year. The breed must also be currently held by at least five different breeders in different locations throughout the nation. Breeds termed "Critical" are the Andalusian, Aseel, Buckeye, Buttercup, Campine, Catalana, Chantecler, Crevecouer, Delaware, Faverolle, Holland, Houdan, La Fleche, Malay, Redcap, Spanish, Nankin, Russian Orloff, and Sumatra.

Breeds termed "Threatened" are the Ancona, Cubalaya, Dorking, Java, Lakenvelder, Langshan, and Sussex.

Breeds in the "Watch" category are the Brahma, Cochin, Cornish (non-industrial), Dominique, Hamburg, Jersey Giant, Minorca, New Hampshire, Polish, and Rhode Island White.

Breeds termed "Recovering" are the Australorp, Leghorn (non-industrial), Orpington, Plymouth Rock (non-industrial), Rhode Island Red, and Wyandotte.

Critical breeds are globally threatened and exist in the United States in known populations of five hundred or fewer birds. Five or fewer primary breeding flocks of fifty or more birds are known to exist in the country.

Threatened breeds are also globally at peril and exist in America in a known population of one thousand breeding birds are less. They are being held in no more than ten primary breeding flocks of fifty birds or more.

Birds in the "Watch" grouping have rather limited geographic distribution and have a breeding population of five thousand or fewer birds in this county. They are backed by ten or fewer primary breeding flocks.

Recovering breeds have exceeded the numbers for inclusion in the Watch category, but still merit close monitoring and increased support.

Large fowl breeds and varieties held by reporting members of the Society for the Preservation of Poultry Antiquities in their Directory (not all breeds are currently sanctioned by the American Poultry Association):

Barred Ameraucana

Black Ameraucana

Black Breasted Red Ameraucana

Blue Ameraucana

Blue Wheaten Ameraucana

Buff Ameraucana

Silver Ameraucana

Silver Duckwing Ameraucana

Wheaten Ameraucana

White Ameraucana

Muffed American Game

Red Quill American Game

Rosecomb Ancona

Single Comb Ancona

Blue Andalusian

Spitzhauben Appenzeller

Gold Spitzhauben Appenzeller

Black Araucana

Black Breasted Red Araucana

Blue Araucana

Silver Araucana

White Araucana

Black Aseel

Black Breasted Red Aseel

Dark Aseel

Hydrobad Aseel

Pakistani Aseel

Pumpkin Aseel

Pyle Aseel

Red Aseel

Spangled Aseel

Wheaten Aseel

White Aseel

Australorp

Barnevelder

Gold Spangled Brabanter

Buff Brahma

Dark Brahma

Light Brahma

Partridge Brahma

Buckeye Buttercup

Golden Campine

Silver Campine

Buff Catalana

Black Chantecler

Buff Chantecler

Partridge Chantecler

White Chantecler

Barred Cochin

Birchen Cochin

Black Cochin

Blue Cochin

Brown Red Cochin

Buff Cochin

Golden Laced Cochin

Partridge Cochin

Red Cochin

Red Frizzle Cochin

Silver Laced Cochin

Splash Cochin

White Cochin

Buff Cornish

Dark Cornish

White Cornish

White Laced Red Cornish

Crevecoeur

Black Breasted Red Cubalaya

Light Gray Cubalaya

Golden Duckwing Cubalaya
Red Pyle Cubalaya
Silver Duckwing Cubalaya
Delaware
Dominique
Single Comb Black Dorking
Blue Breasted Red Dorking
Single Comb Brown Red Dorking
Buff Dorking
Rosecomb Colored Dorking
Single Comb Colored Dorking
Rosecomb Crele Dorking
Single Comb Crele Dorking
Rosecomb Cuckoo Dorking
Single Comb Cuckoo Dorking
Dark Birchen Gray Dorking
Light Gray Dorking
Single Comb Red Dorking
Rosecomb Red Dorking
Red Pyle Dorking
Silver Blue Dorking
Single Comb Silver Gray Dorking
Spangled Dorking
White Dorking
Yellow Dorking
Silver Egyptian Fayoumis
Cuckoo Faverolle
Salmon Faverolle
Mahogany Faverolle
Flathead Game
Blue Hamburg
Golden Penciled Hamburg
Golden Spangled Hamburg
Silver Penciled Hamburg

Silver Spangled Hamburg
Barred Holland
Mottled Houdan
White Houdan
Icelandic
Indian Game
Iowa Blue
Black Java
Mottled Java
White Java
Black Jersey Giant
Blue Jersey Giant
White Jersey Giant
Red Jungle Fowl
Black Breasted Red Kraienkoppe
Golden Kraienkoppe
Silver Kraienkoppe
White Kraienkoppe
Black La Fleche
Blue La Fleche
White La Fleche
Lakenvelder
Silver Lakenvelder
Golden Lakenvelder
Birchen Langshan
Black Langshan
Blue Langshan
White Langshan
Layorca
Rosecomb Black Leghorn
Single Comb Black Leghorn
Single Comb Black Tailed Red
   Leghorn
Rosecomb Buff Leghorn

Single Comb Buff Leghorn
Single Comb Columbian Leghorn
Rosecomb Dark Brown Leghorn
Single Comb Dark Brown
    Leghorn
Single Comb Exchequer Leghorn
Golden Leghorn
Single Comb Golden Duckwing
    Leghorn
Rosecomb Light Brown Leghorn
Single Comb Light Brown
    Leghorn
Single Comb Red Leghorn
Rosecomb Silver Leghorn
Single Comb Silver Leghorn
Rosecomb White Leghorn
Single Comb White Leghorn
Black Long Crower
Madagascar Game
Black Breasted Red Malay
Pyle Malay
White Malay
Manx Rumpless
Black Marans
Crele Marans
Cuckoo Marans
Gold Cuckoo Marans
Rosecomb Black Minorca
Single Comb Black Minorca
Single Comb Buff Minorca
Rosecomb White Minorca
Single Comb White Minorca
Birchen Modern Game
Black Breasted Red Modern
    Game

Brown Red Modern Game
Lemon Blue Modern Game
Red Pyle Modern Game
Silver Blue Modern Game
Black Naked Neck
Buff Naked Neck
Red Naked Neck
New Hampshire Red
Norwegian Jaerhon
Black Breasted Red Old English
    Game
Black Breasted Red Old English
    Game (Muffed)
Black Breasted Red Old English
    Game (Tasseled)
Blue Breasted Red Old English
    Game
Blue Breasted Red Old English
    Game (Muffed)
Brown Red Old English Game
Golden Duckwing Old English
    Game
Red Pyle Old English Game
Spangled Old English Game
White Old English Game
Spangled Morgan White Hackle
    Old English Game
Buff Orloff
Crele Orloff
Cuckoo Orloff
Mahogany Orloff
Spangled Orloff
White Orloff
Black Orpington
Blue Orpington

Buff Orpington
Splash Orpington
White Orpington
Black Penedesenca
Orele Penedesenca
Dun Phoenix
Ginger Phoenix
Silver Phoenix
Golden Phoenix
Barred Plymouth Rock
Black Plymouth Rock
Buff Plymouth Rock
Columbian Plymouth Rock
Golden Laced Plymouth Rock
Partridge Plymouth Rock
Silver Penciled Plymouth Rock
White Plymouth Rock
Bearded Black Polish
Black Polish
Black Crested Blue Polish
Black Crested White Polish
Blue Polish
Bearded Buff Laced Polish
Buff Laced Polish
Cuckoo Polish
Frizzle Polish
Bearded Golden Polish
Golden Polish
Bearded Silver Polish
Silver Polish
Tolbunt Polish
Bearded White Polish
White Crested Black Polish

Bearded White Crested Black
   Polish
White Crested Blue Polish
White Crested Chocolate Polish
Redcap
Rosecomb Rhode Island Red
Single Comb Rhode Island Red
Rosecomb Rhode Island White
Single Comb Rhode Island White
Rumpless Game
Black Breasted Red Saipan
Wheaten Saipan
White Saipan
Black Shamo
Black Breasted Red Shamo
Blue Shamo
Blue Crested Red Shamo
Dark Shamo
Red Pyle Shamo
White Shamo
Spanish Game
White Faced Black Spanish
White Faced Blue Spanish
White Sultan
Black Frizzle Sumatra
Blue Sumatra
Dun Sumatra
Silver Sumatra
White Sumatra
Buff Sussex
Light Sussex
Speckled Sussex
Red Sussex

Toumaru

Welsummer

Black Wyandotte

Blue Wyandotte

Blue Laced Red Wyandotte

Buff Wyandotte

Columbian Wyandotte

Golden Laced Wyandotte

Silver Laced Wyandotte

Silver Penciled Wyandotte

White Wyandotte

Red Shouldered Yokohama

White Yokohama.

View this as a wish list more than a shopping list. Some of these may be regional variants of the same breed, just one or two people hold some, some are creative works still in progress, and some have not been recognized by the American Poultry Association.

Abroad there are available even more varieties of some of these breeds. Outside of North America some of these breeds may even be known by different names. In England the Cornish is called the Jubilee and is bred in a near score of colors and patterns. I have recently seen photos of Wyandottes in a great many more colors and patterns than we have here including one or two with blue points and a Buff Columbian variety. A Partridge Wyandotte variety is even beginning to be seen here now.

Also from the Directory of the Society for the Preservation of Poultry Antiquities is this list of large fowl breeds known from group records, but not recently documented.

Now here is a wish list as a couple of these may even be extinct! Anyone with a few extra Violet Laced Wyandottes, please feel free to give me a call.

A reading of the poultry literature will reveal birds that didn't make this list including Red and Buff Barred Plymouth Rocks.

Brown Brabenter

Cream Brabanter

Breda

Breese

Black Cornish

Blue Cornish

White Coveney

Black Cubalaya

Blue Crested Red Cubalaya

Blue Golden Duckwing Cubalaya

Wheaten Cubalaya

White Cubalaya

Gold Fayoumi

Essex

Black Hamburg

Silver Laced Hamburg

White Hamburg

White Holland

Hungarian Yellow

Ixworth

Kimi

Lamona

Barred Leghorn

Rosecomb Mille Fleur Leghorn

Single Comb Mille Fleur Leghorn

White Frizzle Leghorn

Blue Breasted Red Malay

Black Malay

Malines

Black Copper Maran

Marsh Daisy

Rosecomb Buff Minorca

Black Modern Game

Blue Modern Game

Golden Duckwing Modern Game

Silver Duckwing Modern Game

White Modern Game

Norfolk Gray

North Holland Blue

Black Old English Game

Brown Red Old English Game

Blue Old English Game

Blue Golden Duckwing Old
   English Game

Crele Old English Game

Ginger Red Old English Game

Ginger Red Old English Game

Gray Old English Game

Henny Old English Game

Lemon Blue Old English Game

Silver Duckwing Old English Game

Wheaten Old English Game

Old English Pheasant Fowl

Black Orloff

Brown Red Orloff

Black Breasted Red Orloff

Mottled Orloff

Red Orloff

Persian Rumpless

Black Phoenix

Black Breasted Red Phoenix

Blue Phoenix

Blue Golden Phoenix

Brown Red Phoenix

Lemon Blue Phoenix

Quail Phoenix

White Phoenix

Blue Plymouth Rock

Rheinlander

Scots Dumpy

Scots Grey

Blue Wheaten Shamo

Spangled Shamo

Wheaten Shamo

Black Sultan

Blue Sultan

Thuringer

Tozo

Vorwerk

White Surrey

Wilmslow

Barred Wyandotte

Blue Laced Wyandotte

Blue Laced Golden Wyandotte

Buff Laced Wyandotte

Red Wyandotte

White Laced Buff Wyandotte

Violet Laced Wyandotte

Black Wyndham

Black Breasted Red Yokohama

Golden Duckwing Yokohama

Light Brown Yokohama

Silver Duckwing Yokohama

Spangled Yokohama

York Fowl

POULTRY ASSOCIATIONS

American Poultry Association; PO Box 306, Burgettstown, PA 15021, phone: 724-729-3459, email: secretaryapa@yahoo.com, website: www.amerpoultryassn.com.

American Livestock Breeds Conservancy; PO Box 477, Pittsboro, NC 27312, phone: 919-542-5704, fax: 919-545-0022, email: albc@albc-usa.org, website: www.albc-usa.org.

Society for the Preservation of Poultry Antiquities; 1057 Nick Watts Rd, Lugoff, SC 29078, phone: 570-837-3157, email: christine.heinrichs@gmail.com, website: www.feathersite.com/Poultry/SPPA/SPPA.html.

POULTRY CLUBS

Ameraucana Breeders Club; 33878 Highway 87, California, MO 65018, phone: 573-796-3999, email: Michael@bantamhill.com, website: www.ameraucana.org/index.html.

American Bantam Association; PO Box 127, Augusta, NJ 07822, phone: 973-383-8633, email: fancybantams@embarqmail.com, website: www.bantamclub.com.

American Brahma Club; e-mail: henshaven@iclub.org, website: www.americanbrahmaclub.webs.com.

American Brown Leghorn Club; PO Box 602, Stanwood, Washington 98292, phone: 360-629-3356, email: ablc@the-coop.org, website: http://the-coop.org/leghorn/ablc1.html.

American Buttercup Club; 7257 W 48 Road, Cadillac, MI 49601, phone: 231-862-3671, email: americanbuttecupclub@yahoo.com, website: www.geocities.com/americanbuttercupclub/#menu.

American Dutch Bantam Society; 1910 Union St, Alameda, Ca 94501, email: krista@fynbosfarmpoultry.com, website: www.dutchbantamsocietyamerica.com.

Bearded Belgian d'Anver Club; 11709 Cedar Ridge Rd, Williamsport, MD 21795, Phone: 301-223-6617, email: free2crow@myactv.net, website: http://danverclub.blogspot.com.

American Game Fowl Society; Box 800, Belton, SC 29627, phone: 864-237-5280, email: gamefowls4all@yahoo.com, website: http://americangamefowl.org.

Canadian Araucana Society; 3102 Shawnigan Lake Road, Cobble Hill, BC V0R 1L6, email: canadianaraucanasociety@shaw.ca, website: http://members.shaw.ca/CanadianAraucanaSociety.

Dominique Club of America; 948 West Bear Swamp Road, Walhalla, SC 29691, phone: 864-638-5650, email: bryan_k_oliver@yahoo.com, website: www.dominiqueclub.org.

National Jersey Giant Club; 28143 County Road 4, Pequot Lakes, MN 56472, website: http://nationaljerseygiantclub.com.

North American Hamburg Society; 95393 Grimes Road, Junction City, OR 97448, phone: 541-998-3944, email: info@northamericanhamburgs.com, website: www.northamericanhamburgs.com.

Plymouth Rock Fanciers Club; 14390 South Blvd, Silverhill, AL 36576, email: katz@gulftel.com, website: www.crohio.com/rockclub.

Rhode Island Red Club of America; 28144 Chaparrel Avenue, Taft, CA 93268-9757, phone: 661-765-4804, website: www.crohio.com/reds.

Rosecomb Bantam Federation; PO Box 126, Portales, NM 88130, phone: 575-359-1074, email: firemannm@msn.com, website: www.rosecomb.com/federation.

United Orpington Club; 201 English Mt. Rd. Newport, TN 37821, phone: 423-625-3191, email: unitedorpclub@aol.com, website: www.unitedorpingtonclub.com.

PERIODICALS WITH POULTRY THEMES

Acres U.S.A.; PO Box 91299, Austin, TX 78709, phone: 512-892-4400, fax: 512-892-4448, email: info@acresusa.com, website: www.acresusa.com.

Backyard Poultry Magazine; 145 Industrial Drive, Medford, WI 54451, phone: 715-785-7979 fax: 715-785-7414, email: byp@tds.net, website: www.backyardpoultrymag.com.

Feather Fancier Newspaper; 5739 Telfer Road, Sarnia, ON, Canada N7T 7H2, phone: 519-542-6859, fax: 519-542-4168, email: featherfancier@ebtech.net, website: http://www.featherfancier.on.ca.

Small Farm Today; 3903 West Ridge Trail Road, Clark, MO 65243, phone: 573-687-3525, email: smallfarm@socket.net, website: www. smallfarmtoday.com.

Poultry Press Newspaper; PO Box 542; Connersville, IN 47331, phone: 765-827-0932, fax: 765-827-4186, email: info@poultrypress. com, website: www.poultrypress.com.

A LIST OF SMALL HATCHERIES & POULTRY SUPPLY FIRMS:

Braggs Mountain Poultry; 1558 Kreider Road, Ft. Gibson, OK 74454, phone: 254-622-8169, website: www.braggsmountainpoultry.com. Purveyors of Braggs Mountain Buffs, a breed good for both meat and large brown eggs.

Cackle Hatchery; PO Box 529, Lebanon, MO 65536, phone: 417-532-4581, fax: 417-588-1918, email: cacklehatchery@cacklehatchery. com , website: www.cacklehatchery.com. Offering 181 varieties of egg and meat poultry breeds, rare breeds, turkeys, ducks, geese, guinea, game, and peafowl; plus equipment and supplies.

Clearview Stock Farm & Hatchery; Box 399, Gratz, PA 17030, phone: 717-365-3234.

C.M. Estes Hatchery; PO Box 5776, Springfield, MO, phone: 417-865-8874, fax: 417-865-2242, website: www.esteshatchery.com. Supplying rare pheasants, quail, chukar, common poultry breeds and some heritage poultry, equipment and supplies.

Cutler's Pheasant and Poultry Supply; 1940 Old 51, Applegate, MI 48401, phone: 810-633-9450, fax: 810-633-9178, email: sales@ cutlersupply.com, website: www.cutlersupply.com. Offering supplies for pheasant, poultry and other fowl rearing, as well as netting, leg bands, medicines, egg trays, incubators, feeders, waterers, brooders, chicken nests and more.

Double R Discount Supply; 5156 Minton Road NW, Palm Bay, FL 32907, phone: 321-837-1625, fax: 321-837-1628, email: customerservice@dblrsupply.com, website: http://dblrsupply.com.

Offering live chicks, ducklings, goslings, hatching eggs and supplies such as incubators, hatchers, and cages.

Dunlap Hatchery; PO Box 507, Caldwell, ID 83606, phone: 208-459-9088, fax: 208-455-0665, website: www.dunlaphatchery.net. Shipping broiler, standard, exotic, bantam, duck, turkey, geese, and gamebird chicks. Also supplying feeders, waterers, incubators, leg bands and medications.

Eagle Nest Poultry; Box 504, Oceola, OH 44860, phone: 419-562-1993, website: www.eaglenestpoultry.com. Hatchery featuring chicks and hatching eggs for old-fashioned breeds, waterfowl, turkey, peafowls, and gamebirds.

Eggcartons.com; 9 Main Street, Suite 1F, PO Box 302, Manchaug, MA 01526, phone: 888-852-5340, fax: 877-455-4647, email: email@ eggcartons.com, website: www.eggcartons.com. Selling egg cartons, egg trays, incubators, feeders, waterers, laying nests, egg washers, chicken tractors, brooders, shipping boxes and more.

First State Veterinary Supply; PO Box 190, Parsonburg, MD 21849, phone: 410-546-6137, website: www.firststatevetsupply.com. A business operated by the chicken doctor, Peter Brown and offering medicines, vitamins and wormers for poultry. Also offering feeders, waterers, leg bands and crates.

Foy's Pigeon Supplies; 3185 Bennett's Run Road, Beaver Falls, PA 15010, phone: 877-355-7727, fax: 724-843-6070, email: foyspigeon@zoominternet.net, website: www.foyspigeonsupplies.com. Offering leg bands, drinkers, medicines, wormers, coops and cages. Specializing in racing homer pigeons.

Heartland Hatchery; Rt. 1 Box 177-A, Amsterdam, MO 64723, phone: 660-267-3679, email: jnieder@ckt.net, website: www. heartlandhatchery.com. Offering fertile hatching eggs for standard chickens, rare breed poultry, bantam chickens, fancy breed turkeys, ducks, goslings and guineas.

Hoffman Hatchery; P.O. Box 129V, Gratz, PA 17030, phone: 717-365-3694, email: info@hoffmanhatchery.com, website: www. hoffmanhatchery.com. A hatchery offering goslings, ducklings, commercial breed chicks, turkey poults, guineas, game birds, bantams and fancy poultry. Also supplying brooders, fountains, feeders, leg bands, egg nests, egg cartons, and incubators.

Hoover's Hatchery; 205 Chickasaw St, PO Box 200, Rudd, IA 50471, phone: 800-247-7014, fax: 641-395-2208, email: hoovers@omnitelcom.com, website: http://hoovershatchery.com. Offering white and brown egg layering, broilers, bantam and colored egg laying chicks. Also supplying duck, geese, pheasant, turkey, peafowl and guinea chicks. Supplies also include fountains, feeders, incubator, leg bands and vitamins.

Ideal Poultry Breeding Farms; PO Box 591, Cameron, TX 76520, phone: 254-697-6677, fax: 254-697-2393, email: help@idealpoultry.com, website: www.ideal-poultry.com. Supplying hatching eggs and chicks of rare white and brown egg layers, broilers, ducks, chukar, turkeys and bantams.

Lazy 54 Farm; PO Box 429, Hubbard, OR 97032, phone: 503-981-7801, fax: 503-981-7215, email: hatchery@earthlink.net, website: www.shankshatchery.com. Offering chicks of large fowl production birds, bantam, turkey, ducks, geese, wild birds, ringneck pheasants, chukars and Bobwhite quail.

Marti Poultry Farm; PO Box 27, Windsor, MO 65360, phone: 660-647-3156, fax: 660-647-3999, email: sales@martipoultry.com, website: www.martipoultry.com. Hatching over 100 varieties of rare, fancy and exotic breeds of chickens, bantams, ducks and guineas.

Meyer Hatchery; 626 State Road 89; Polk, OH 44866, phone: 888-568-9755, fax: 419-945-9841, email: info@meyerhatchery.com, website: www.meyerhatchery.com. Offering poultry supplies, coops, waterers, feed and supplements, incubators, brooder supplies, day-old chicks such as Barred Plymouth Rocks, Black Star, Buckeyes, Delawares, Dominique, and Black Australorps. They also carry turkeys, ducks, game birds, peafowl and geese.

Myers Poultry Farm; 966 Ragers Hill Road, South Fork PA 15956, phone: 814-539-7026, email: myerspf@juno.com. Offering chicks in eight varieties including Cornish crosses, Red Cornish and most of the common layers. Also available are ducks, guineas, geese, turkeys, bobwhites and chukars.

Phinney Hatchery; 1331 Dell Avenue, Walla Walla, WA 99362, phone: 509-525-2602, fax: 509-529-8136. Hatchery supplying rare breed chicks, ringneck pheasants, chukar, ducklings and goslings.

Poultryman's Supply Company; PO Box 612, Columbus, IN 47201, phone: 812-603-7722, email: info@poultrymansupply.com, website: www.poultrymansupply.com. Poultry supplies including egg cartons, processing equipment, snap clamps, medications, feeders, waterers, leg bands, brooders and incubators.

Privett Hatchery; PO Box 176; Portales, NM 88130, phone: 575-356-6425, fax: 575-356-6540, email: privetth@yahoo.com, website: www.privetthatchery.com. A hatchery that supplies chickens, bantams, ducks, geese and guineas.

Reich Poultry Farms; 1625 River Road, Marietta, PA 16547, phone: 717-426-3411, fax: 717-426-8061, email: info@reichpoultryfarm.com, website: http://reichspoultryfarm.com. Hatchery and shipper of eight varieties of day-old baby chickens such as Rhode Island Reds and Barred Rocks, turkeys, geese and guinea hens.

Ridgway Hatchery; P.O. Box 306; LaRue, OH 43332, phone: 740-499-2163, fax: 740-499-2828, email: RidgwayEgg@aol.com, website: www.ridgwayhatchery.com. A breeder and hatchery supplying Rhode Island Red, New Hampshire, Buff Orpington, Barred Rock, White Rock, Araucana, Silver Laced Wyandotte chickens, ducks, geese, guineas, game birds and turkeys.

Sand Hill Preservation Center; 1878 230th Street; Calamus, IA 52729, phone: 563-246-2299, email: sandhill@fbcom.net, website: www.sandhillpreservation.com. Breeders of rare poultry, quail, guineas, ducks, geese and turkeys.

Schlecht Hatchery; 9749 500th Avenue; Miles, IA 52062, phone: 563-682-7865, fax: 563-682-7450, email: poultry@schlechthatchery.com, website: www.schlechthatchery.com. Offering geese, ducks, bantam poultry and various large fowl such as Brahmas, Polish, Orpingtons, Reds, Rocks and Wyandottes.

Shady Lane Poultry Farm; PO Box 612, Columbus, IN 47201, phone: 812-603-7722, email: info@shadylanepoultry.com, website: www.shadylanepoultry.com. A breeding farm offering exhibition quality large fowl, heirloom and rare poultry. Also offer bantams in several breeds and varieties.

Smith Poultry Supply; 14000 West 215th Street; Bucyrus, KS 66013, phone: 913-879-2587, fax: 913-533-2497, email: smithkct@ earthlink.net, website: www.poultrysupplies.com. John and Terry Smith offer incubators, brooders, laying nests, show and pastured poultry breeds, pigeons, peafowl, game birds and waterfowl.

Townline Hatchery; PO Box 108, Zeeland, MI 49464, phone: 616-772-6514, fax: 616-772-2969, email: townlinehatchery@sbcglobal.net, website: www.townlinehatchery.com. Offering ducks, turkeys, geese, pheasants, layer and broiler chicks for sale.

Urch/Turnland Poultry; 2142 NW 47 Ave, Owatonna, MN 55060, phone: 507-451-6782. Duane Urch breeds and sells show quality fowl. Eggs and chicks available in many standard and bantam breeds.

LARGER HATCHERIES

Belt Hatchery; 7272 S West Ave, Fresno, CA 93706, phone: 559-264-2090, fax: 559-264-2095, email: orders@belthatchery.com, website: www.belthatchery.com. Offering day old chicks of standard, heavy and specialty poultry breeds, bantams and turkeys.

Moyer's Chicks, 266 E Paletown Rd, Quakertown, PA 18951, phone: 215-536-3155, fax: 215-536-8034, email: orders@moyerschicks. com, website: www.moyerschicks.com. Offering broiler, brown egg layers and white leghorn chicks, pullets and fertile eggs as well as incubators for sale.

Mt. Healthy Hatcher; 9839 Winton Rd, Mt. Healthy, OH 45231, phone: 800-451-5603, fax: 513-521-6902, email: info@mthealthy. com, website: www.mthealthy.com. Supplying 24 breeds and varieties of day old chicks from Buff Orpingtons, Partridge Rocks, Black Australorps, New Hampshires, Speckled Sussex and more. Hatching eggs also available in small quantities. Also supplies ducks, turkeys and gamebirds.

Murray McMurray Hatchery; PO Box 458, 191 Closz Drive, Webster City, IA 50595, phone: 515-832-3280, fax: 515-832-2213, website: www.mcmurrayhatchery.com. Supplying day old chicks in standard breeds and bantams. Also available are partridge, quail, pheasants, ducklings, guineas, peafowl, goslings and turkeys. Supplies for sale include incubators, waterers, feeders, brooders, cages and pens.

Sun Ray Chicks Hatchery; 106 North Main Street, PO Box 300, Hazleton, IA. 50641, phone: 319-636-2244, email: info@ sunrayhatchery.com, website: www.sunrayhatchery.com. Offering jumbo Cornish Rock broilers, various laying pullets, Black Australops, and other colored poultry for ethnic markets.

Stromberg's Chicks and Gamebirds Unlimited; PO Box 400, Pine River, MN 56474, phone: 218-587-2222, fax: 218-587-4230, email: info@strombergschickens.com, website: www.strombergschickens. com. Supplies available include incubators and brooders, feeders and waterers. Birds, chicks and eggs available for large poultry fowl, bantams, ducks, geese, turkeys, guineas, peafowl, pigeons, doves, button quail, pheasants, quail, partridge, wild turkeys and wild waterfowl.

Welp Hatchery; PO Box 77, Bancroft, IA 50517, phone: 1-800-458-4473, fax 515-885-2346, email bkollasch@welphatchery.com, website: www.welphatchery.com. Specializing in Cornish Rock broilers; also supplying egg layer poultry breeds, standard breeds, bantams, quail, ducks, turkeys, goslings, pheasants, goslings, chukar partridges, guineas and rare breeds like Black Wyandotte, Buff Brahma, Cuckoo Maran, Golden Spangled Hamburg, Silver Laced Polish and more.

SUGGESTED FOR FURTHER READING

Backyard Poultry Naturally, Alanna Moore, Acres U.S.A., 2007.

Chicken Coops — 45 Building Plans for Housing Your Flock, Judy Pangman, Storey Publishing, 2006.

Chicken Tractor, Andy Lee and Pat Foreman, Good Earth Publications, 2006.

Genetics of the Fowl — The Classic Guide to Chicken Genetics and Poultry Breeding, Frederick Hutt, Norton Creek Press, 2003.

How to Raise Chickens — Everything you Need to Know, Christine Heinrichs, Voyageur Press, 2007.

How to Raise Poultry, Christine Heinrichs, Voyageur Press, 2009.

Living with Chickens, Jay Rossier, The Lyons Press, 2002.

Pastured Poultry Profits, Joel Salatin, Polyface, Inc, 1999.

Raising Poultry on Pasture: 10 Years of Success, Jody Pangman (editor), American Pastured Poultry Producers, 2006.

Raising Poultry Successfully, Will Graves, Williamson Publishing, 1985.

Storey's Illustrated Guide to Poultry Breeds, Carol Ekarius, Storey Publishing, 2007.

Success with Baby Chicks, Robert Plamondon, Norton Creek Press, 2003.

The Mating and Breeding of Poultry, Harry M. Lamon and Rob R. Slocum, The Lyons Press, 1920 (Orange Judd Company), reprint 2003 (Lyons Press).

POULTRY FORUMS FOR ONLINE SUPPORT

The PoultrySite Discussion Forum: *www.thepoultrysite.com/forums*

The Classroom @ The Coop: *www.the-coop.org/cgi-bin/UBB/ultimatebb.cgi*

BackYardChickens Forum: *www.backyardchickens.com/forum*

Poultry Chat Poultry Forum: *www.poultrychat.com/vb*

The Poultry Keeper Forum (UK): *www.poultrykeeper.com/forum*

TOE PUNCH GUIDE

The numbers one through sixteen can be encoded into a chick via the toe punch system. These punches can denote the hatch from which the chicks came, the breeding pen that produced them, their time of hatch or other designation.

The punches are painless and made in the fleshy webs between the chick's toes. An all-metal toe punch will cost about five dollars. Keep it clean and disinfect it in a simple alcohol bath before and after each use.

Make the punches crisp and quick. Then be sure the small flap of flesh is completely removed to prevent the punch from closing back up.

The following is a suggested sequence to use as a toe-punching guide.

1 = no punch in any web

2 = one punch in the left foot outside web

 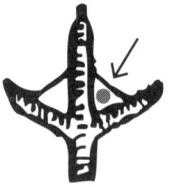

3 = one punch in the right foot outside web

4 = one punch in the right foot inside web

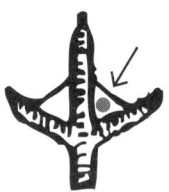 

5 = one punch in the left foot inside web

6 = one punch in the left foot outside web and one punch in the right foot inside web

7 = one punch in the left foot
inside web and one punch in
the right foot inside web

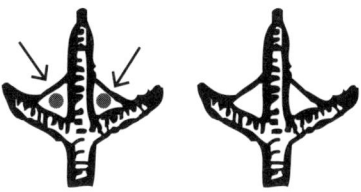

8 = both left foot webs
punched

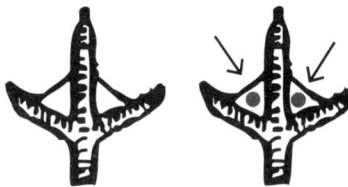

9 = both right foot webs
punched

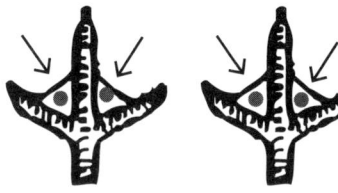

10 = punch both webs on both
feet

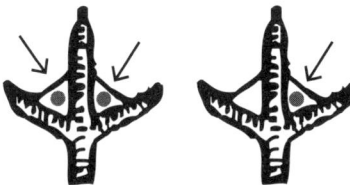

11 = punch both webs on the
left foot and the outside web
on the right foot

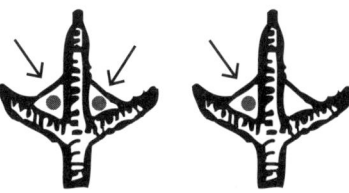

12 = punch both webs on the
left foot and the inside web on
the right foot

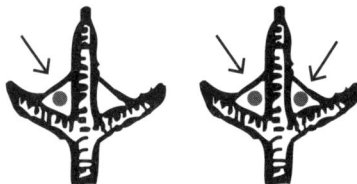

13 = punch both webs on the
right foot and the outside web
on the left foot

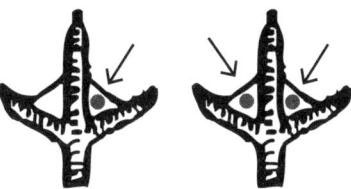

14 = punch both webs on the
right foot and the inside web
on the left foot

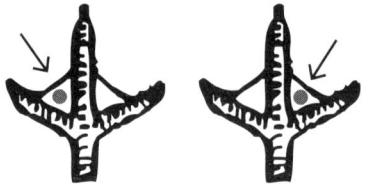 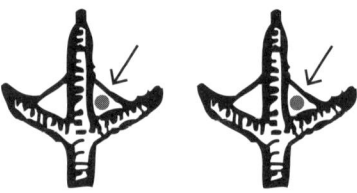

15 = one punch on the
left food outside web and one
punch on the right foot
outside web.

16 = one punch on e the left
foot inside web and one punch
on the right foot outside web

• SINGLE COMB •

• WALNUT COMB •

• V-COMB •

STRAWBERRY
• COMB •

BUTTERCUP
• COMB •

• ROSECOMB •

• PEA COMB •

CUSHION
• COMB •

• INDEX •

Alderson Brothers, 20

Ameraucana, 3

American Bantam Association, 119

American Livestock Breeds Conservancy (ALBC), 7, 63, 65, 71, 166, 168, 179, 195, 212, 311, 326

American Poultry Association, 38, 53, 54, 63-64, 85, 170

Americauna, 87

Ancona, 254, 280-281, 284, 332, 333

Andalusian, 284

Andrews, David, 201, 301, 302-304

Araucana, 86

Audet, Marye, 50

Australorp, 82, 254, 302, 312, 327, 328, 342

Avian Influenza, 107, 197, 300

Babcock, Monroe, 173

bantam, 48, 298

Barnevelder, 7, 85, 86

Berry Hatchery, 21

bio-security, 107

birds, ornamental, 297

birds, sourcing, 326

blue-egg breed, 86

Bourbon Red turkey, 224

Braggs Mountain Poultry, 301, 303-304

Braggs Mountain, 201, 233, 301-304

Brahma, 144, 254, 284, 302, 344, 346

breed choice, 76, 79

breed preservation, 90

breed ranking, 254

breed selection, 87, 327

breed, historic, 343-347

breeders, 89

breeding assessment, 166

breeding pen, 160

breeding standards, 202-204, 205-206

breeding, 23, 55-58, 70, 73-75, 78, 145-152, 156-165

breeding, best to best, 149

breeding, breed type, 170

breeding, color, 170

breeding, disease resistance, 171

breeding, egg production, 207-211

breeding, egg-laying ability, 168

breeding, fertility & vigor, 171

breeding, mature size, 169

breeding, non-sterile environment, 153-156

breeding, rate of growth, 168

breeding, vigor, 172

broiler, 78-79

broiler, marketing, 266-269

brooder ill, 109

brooder, care, 126-127

brooders, hatchers and incubators, 123-133

broody hens, 217, 219-220

brown egg production breeds, 83-86, 327-331

Buckeye, 168, 255

Buttercup, 335

Cackle Hatchery, 237-243

cage-free eggs, 41

Campine, 335

candling, 227-228

capon, 97-98

Catalana, 335

catalog ordering, 92-96

Chantecler, 11, 152, 255, 284

chick sexing, 24, 233-234

chick, hatching, 228-236

chicken tractor, 105, 110

chicks, ordering, 66-67

Cinnamon Queen, 330

Cochin, 72, 101, 116, 218, 246, 254, 284, 322, 338, 346

cold housing, 107

Columbian, 284, 337

Comet, 330

Cornish X, 14, 18, 65, 77, 137-142, 207, 208, 258, 266, 339-340, 342

Cornish, 13, 209-210, 224, 342, 343

culling, 171

Delaware, 49, 82, 254, 336, 337, 341

Dominecker, 31

Dominique, 31, 44, 52, 83, 247, 254

Dorking, 91, 144, 255, 281-282, 284

dual-purpose breeds, 17, 19, 48, 88, 167, 335

dual-purpose poultry, breeding, 174-178

ducks, 298

E. coli, 300

Easter eggers, 87

egg breeds, 4, 284, 286

egg eating, 113

egg producer, 81-82

egg-laying breeds, 41, 82

egg-mobile, 105

eggs, 82

eggs, marketing, 249-252, 261-266

Faverolle, 91, 284

Fayoumis, 335

feather picking, 235-236

feed rations, commercial, 187

feed, 183-191

feed, breeding birds, 190-191

feed, chick, 188-189

feed, storage, 134

feeders, 120-121

Feeds and Feeding, 109

fencing, 110-111

Fiddler on the Roof, 72

first aid, 183-194

flock building, 90

Frankenfood, 18, 341

Frizzle, 116-117

Future Farmers of America (FFA), 101

gamebirds, 296

Gamecock, 13

geese, 298

Giant, 13, 16, 43, 254, 284, 302, 342

Gibson, 290

Gibson, Robert, 278

Golden Ranger, 224, 225

guinea, 297

Haley, Alex, 13

Hamburg, 100, 254, 284, 335

hatchability, 227-228

Hatchery, 36-37

hatching eggs, 36, 58-62, 69, 217, 223, 226-233

hatching eggs, marketing, 269-272

heat lamps, 126

Heirloom Heritage Farms, 212, 214

hen house design, 102-113

hen house, predators of, 114, 120

heritage breeds, 63-66

Hobbs, Dane, 188

Holland, 49, 335

Houdan, 11, 28, 151, 255

Hungarian, 321

hybridization, 17

incubator, still air, 128-129

incubators, 220-223

Jaerhon, 335

Java, 84, 246, 255

John, Matt, 34

Jull, M.A., 53

Klober, Kelly, 260

Knight Chicks, 119

Knight, Jaynese, 119

La Fleche, 335

Label Rouge, 139, 225

Lakenvelder, 335

Lamon, Harry, 320

Lamona, 21

Langshan, 216, 302, 344, 345

Law of Ten, the, 173

layer breeds, 19

Leghorn, 3, 7, 13-14, 18, 21, 29, 32, 43, 48, 52, 65, 83, 88, 158, 172, 204, 231, 254, 258, 284, 302, 330-336, 338 line breeding, 320

line mating, 147

Maran, 12, 40, 83, 85, 86, 247, 254, 344

marketing rare and heritage breeds, 305-306, 308

marketing, 247-252, 261-277

Marquette, Joseph, 278, 282, 290

Mating and Breeding of Poultry, the, 320

mating, line, 147

mating, rolling, 146

McMurray Hatchery, 150

meat bird, 208

meat production breeds, 16, 284, 339-343

Mediterranean breed class, 84

Minorca, 100, 254, 284, 332, 333, 334

Missouri Department of Agriculture, 105

mites, 200

multiuse production breeds, 335-339

Naked Neck, 83, 225, 255, 344, 345

Narragansett turkey, 224

nativized genetics, 311-317

Nature's Harmony Farm, 142, 156, 226

nest box, 112-114

New Hampshire, 82, 88,151, 254, 284, 327, 336, 337, 341

Noll, Henry, 195-196

Noll's Poultry Farm, 195-196

organic products, 251-252

Orloff, 91

Orpington, 32, 83, 161, 204-205, 219, 232, 254, 284, 292, 302, 322, 323, 327, 329, 336, 338

outcross, 90, 148

parasites, 199-200

peafowl, 297

Penedesenca, 82, 85

Performance Red, 80, 84

pheasant, 297

pigeon, 296

Pit Games, 13

Plymouth Rock, 2, 33, 48, 82, 83, 88, 159, 182, 195, 204, 210, 224, 254, 284, 302, 312, 327, 328, 336, 338, 341

Polish, 28, 38, 210, 255, 344

Polyface Farm, 312, 317

Poulet Rouge, 137, 139 - 141, 224-226

poultry management, 135-136

poultry, basic health care, 201

poultry, confinement, 197

poultry, equipment, used, 133

predators, hen house, 114, 120

production birds, 258

production reds, 80, 84

Promised Land Family Farm, 8-11

pullet, 24

pullet, marketing, 275

quail, 299

range bird, 208

range broiler, 3

range production, 110, 198

rare breeds, 7

Redcap, 335

Rhode Island, 11-12, 18, 32-33, 48, 55, 82, 83, 85, 88, 90, 123, 146, 210, 232, 248, 254, 292, 302, 312, 322, 323-325, 327, 329, 341

Rines, James P., 172

River Hills Poultry Alliance, 253

roaster, 97

Rock, see Plymouth Rock

rolling mating, 146

Roots: The Saga of an American Family, 13

Salatin, Joel, 317

Schrider, Don, 179

Sebright, 216

seedstock, 30, 43, 77, 89, 248

Sex-link, 31, 80, 84, 87, 254, 258, 312, 330

Shady Lane Poultry Farm, 34

Silkie, 9, 118, 218, 247, 255, 298

Sizzle, 115, 116

Slocum, Rob R., 320

Smith, Clifford and Leona, 237

Smith, Jeff, 238, 240, 241, 243

Society for the Preservation of Poultry Antiquities, (SPPA), 7, 71, 279, 282, 294, 286, 334

Spanish, 284, 335

Spitzhauben, 335

Standard of Perfection, 38, 150, 170, 202

string men, 41

Sumatra, 255

Sussex, 284

swan, 297

toe punch, 162-163

Turbyfill, Dawn, 212, 214

Tyson Foods, 76

Tyson, Don, 76

Wallace, Maurice, 56

water, 191-193

waterers, 121-123

Welsummer, 40, 53, 83, 85, 86, 161, 230, 247, 254, 327, 329

Weppner, Karen, 8-11

white egg production breeds, 331-335

Wilder, Laura Ingalls, 20

wing band, 162

winter productivity, 183-184

worms, 199

Wyandotte, 2, 13, 19, 31, 36, 37, 49, 53, 83, 84, 88, 91, 182, 204, 218, 219, 235, 247, 254, 284, 322, 336, 337, 341, 344, 345

Yellow House Farm, 278, 280-282, 290

Young, Liz, 156, 226

Young, Tim & Liz, 142

NATURAL CATTLE CARE

Pat Coleby. Natural Cattle Care encompasses every facet of farm management, from the mineral components of the soils cattle graze over, to issues of fencing, shelter and feed regimens. Coleby presents a comprehensive analysis of farming techniques that keep the health of the animal in mind. She brings a wealth of animal husbandry experiences to bear in this analysis of the serious problems of contemporary farming practices, focusing on how poor soils lead to mineral-deficient plants and ailing farm animals. Coleby provides system-level solutions and specific remedies for optimizing cattle health and productivity. *Softcover, 198 pages.*

NATURAL HORSE CARE

Pat Coleby. Proper horse care begins with good nutrition practices. Chances are, if a horse needs medical attention, the causes can be traced to poor feeding practices, nutrient-deficient feed, bad farming and, ultimately, imbalanced, demineralized soil. Pat Coleby shares decades of experience working with a variety of horses. She explains how conventional farming and husbandry practices compromise livestock health, resulting in problems that standard veterinary techniques can't properly address. *Natural Horse Care* addresses a broad spectrum of comprehensive health care, detailing dozens of horse ailments, discussing their origins, and offering proven, natural treatments. *Softcover, 164 pages.*

NATURAL GOAT CARE

Pat Coleby. Goats thrive on fully organic natural care. As natural browsers, they have higher mineral requirements than other domestic animals, so diet is a critical element to maintaining optimal livestock health. In *Natural Goat Care*, consultant Pat Coleby shows how to solve health problems both with natural herbs and medicines and the ultimate cure, bringing the soil into healthy balance. Topics include: correct housing and farming methods; choosing the right livestock; diagnosing health problems; nutritional requirements and feeding practices; vitamins and herbal, homeopathic and natural remedies; psychological needs of goats; breeds and breeding techniques. *Softcover, 374 pages.*

NATURAL SHEEP CARE

Pat Coleby. A comprehensive guide for all breeders of sheep, whether for wool, meat or milk Coleby draws on decades of experience in natural animal husbandry to provide essential information for both organic and conventional farmers. This edition has been expanded significantly in the areas of breeding for finer wool and meat, land management, sheep management and treatment of health problems. Coleby covers breeds of sheep, wool, meat and milk production, feeding requirements, poisonous plants, land management, minerals and vitamins, herbal, homeopathic and natural remedies, and more. *Softcover, 232 pages.*